Everyday Mathematics®

Student Math Journal 2

The University of Chicago
School Mathematics Project

Mc Graw Hill **Wright Group**

The *McGraw·Hill* Companies

UCSMP Elementary Materials Component

Max Bell, Director

Authors

Max Bell
Jean Bell
John Bretzlauf*
Amy Dillard*
Robert Hartfield
Andy Isaacs*
James McBride, Director
Kathleen Pitvorec*
Peter Saecker

Technical Art

Diana Barrie*

Second Edition only

Photo Credits

Phil Martin/Photography, Jack Demuth/Photography, Cover Credits: Leaf, tomato, buttons/Bill Burlingham Photography, Tree rings background/Peter Samuels/Stone, Photo Collage: Herman Adler Design Group

Contributors

Librada Acosta, Carol Arkin, Robert Balfanz, Sharlean Brooks, Jean Callahan, Anne Coglianese, Ellen Dairyko, Tresea Felder, James Flanders, Dorothy Freedman, Rita Gronbach, Deborah Arron Leslie, William D. Pattison, La Donna Pitts, Danette Riehle, Marie Schilling, Robert Strang, Sadako Tengan, Therese Wasik, Leeann Wille, Michael Wilson

Permissions

page 187: *North American Indian Stickers* by Madeleine Orban-Szontagh. Dover Publications, Inc.

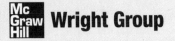

Send all inquiries to:
Wright Group/McGraw-Hill
P.O. Box 812960
Chicago, IL 60681

Printed in the United States of America.

ISBN 0-07-584442-7

12 13 14 15 DBH 10 09 08 07 06

The McGraw·Hill Companies

Contents

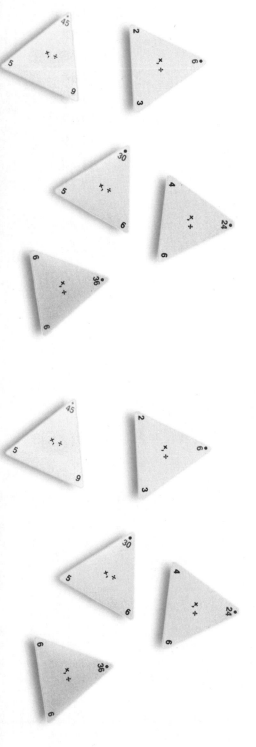

Unit 6: Developing Fact Power

Unit 7: Geometry and Attributes

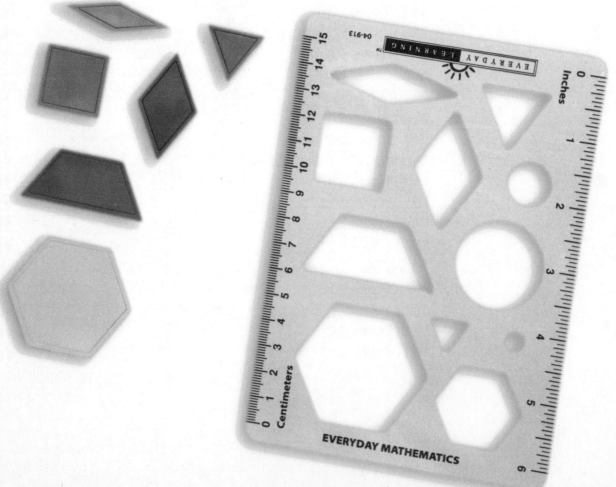

Unit 8: Mental Arithmetic, Money, and Fractions

Unit 9: Place Value and Fractions

Unit 10: End-of-Year Reviews and Assessments

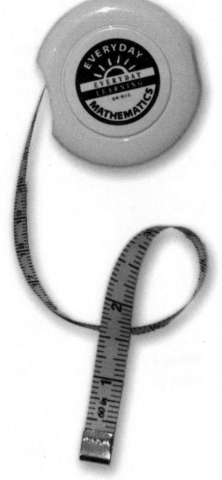

Activity Sheets

Dice-Throw Record 2

Unit

dice dots

Record each fact and its turn-around fact once.

										12
										11
										10
									6 + 3	9
										8
										7
										6
								1 + 4	4 + 1	5
										4
										3
										2

Math Boxes 6.1

1. Add.

$$3 + 1 = \underline{}$$

$$\begin{array}{r} 6 \\ + \ 6 \\ \hline \end{array}$$

$$\begin{array}{r} 8 \\ + \ 8 \\ \hline \end{array}$$

$$\underline{} = 0 + 8$$

2. Find the rule. Fill in the missing numbers.

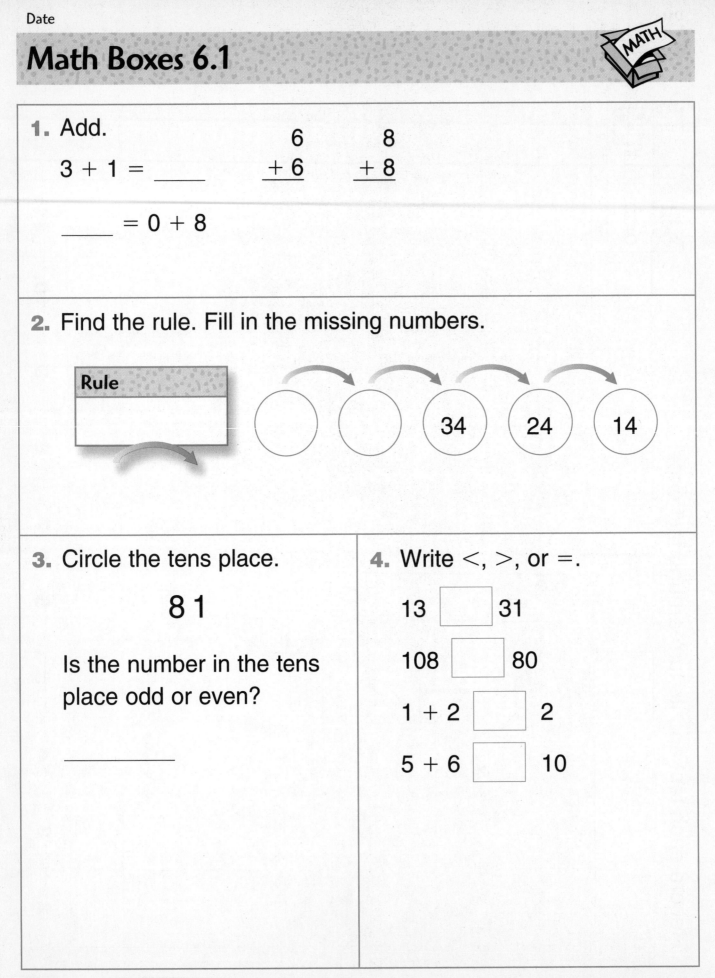

Rule

⟶ ⟶ ⟶ ⟶

◯ ◯ 34 24 14

3. Circle the tens place.

8 1

Is the number in the tens place odd or even?

_____

4. Write $<$, $>$, or $=$.

13 ☐ 31

108 ☐ 80

1 + 2 ☐ 2

5 + 6 ☐ 10

Name-Collection Boxes

1. Write other names for 11.

2. Write other names for 12.

11
8 + 3
13 − 2

12
1 dozen
~~HHT~~ ~~HHT~~ //
3 + 3 + 3 + 3
15 − 3

3. Cross out the names that don't belong in the 10-box.

4. Your choice

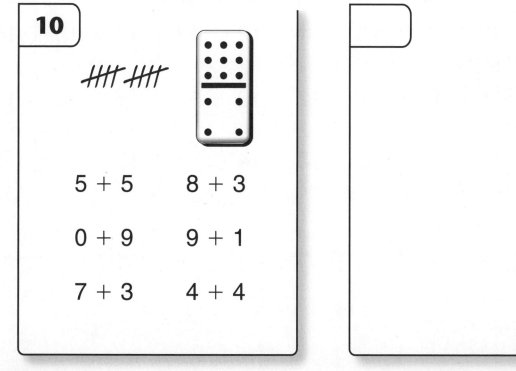

10
~~HHT~~ ~~HHT~~
5 + 5   8 + 3
0 + 9   9 + 1
7 + 3   4 + 4

Addition/Subtraction Facts Table

+,−	0	1	2	3	4	5	6	7	8	9
0	0	1	2	3	4	5	6	7	8	9
1	1	2	3	4	5	6	7	8	9	10
2	2	3	4	5	6	7	8	9	10	11
3	3	4	5	6	7	8	9	10	11	12
4	4	5	6	7	8	9	10	11	12	13
5	5	6	7	8	9	10	11	12	13	14
6	6	7	8	9	10	11	12	13	14	15
7	7	8	9	10	11	12	13	14	15	16
8	8	9	10	11	12	13	14	15	16	17
9	9	10	11	12	13	14	15	16	17	18

Use with Lesson 6.1.

Using the Addition/Subtraction Facts Table 1

1. 5 + 7 = ____   **2.** ____ = 3 + 8   **3.** ____ = 6 + 6

4.   6   **5.**   9   **6.**   7
      + 7   + 6   + 4

7. 9 + 5 = ____   **8.** 8 + 5 = ____   **9.** 8 + 8 = ____

10. 4 + 7 = ____   **11.** ____ = 9 + 7   **12.** ____ = 3 + 9

Complete the "What's My Rule?" tables.

13.

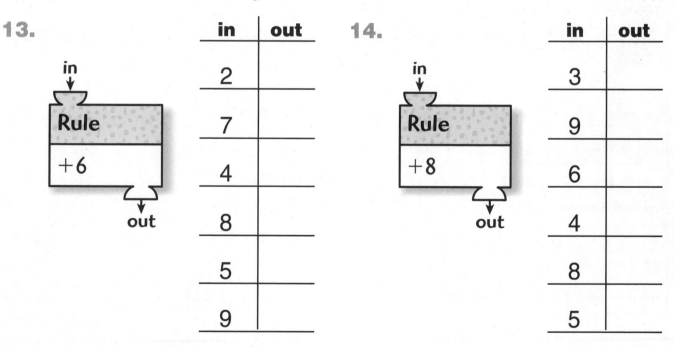

in	out
2	
7	
4	
8	
5	
9	

14.

in	out
3	
9	
6	
4	
8	
5	

Date

1. Draw the hands.

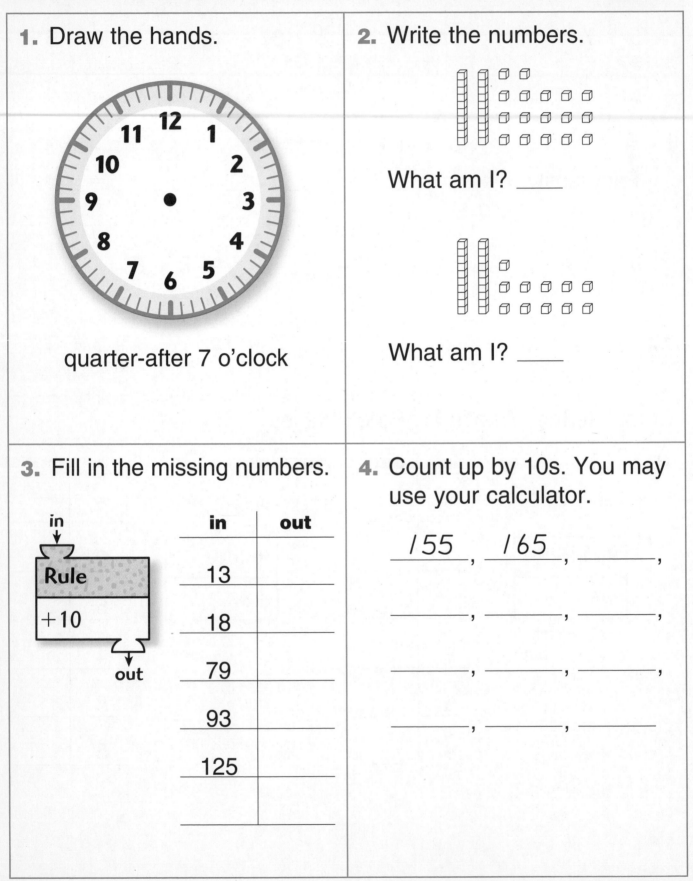

quarter-after 7 o'clock

2. Write the numbers.

What am I? _____

What am I? _____

3. Fill in the missing numbers.

in
↓
Rule

+10

out

in	out
13	
18	
79	
93	
125	

4. Count up by 10s. You may use your calculator.

_155_, _165_, _____,

_____, _____, _____,

_____, _____, _____,

_____, _____, _____

Fact Families

Write the 3 numbers for each domino.
Use the numbers to write the fact family.

1. Numbers: _____, _____, _____

   Fact family: _____ + _____ = _____   _____ − _____ = _____

   _____ + _____ = _____   _____ − _____ = _____

2. Numbers: _____, _____, _____

   Fact family: _____ + _____ = _____   _____ − _____ = _____

   _____ + _____ = _____   _____ − _____ = _____

3. Numbers: _____, _____, _____

   Fact family: _____ + _____ = _____   _____ − _____ = _____

   _____ + _____ = _____   _____ − _____ = _____

4. Make up your own domino. Draw the dots.

   Numbers: _____, _____, _____

   Fact family: _____ + _____ = _____   _____ − _____ = _____

   _____ + _____ = _____   _____ − _____ = _____

Date

MATH

1. Fill in the blanks.

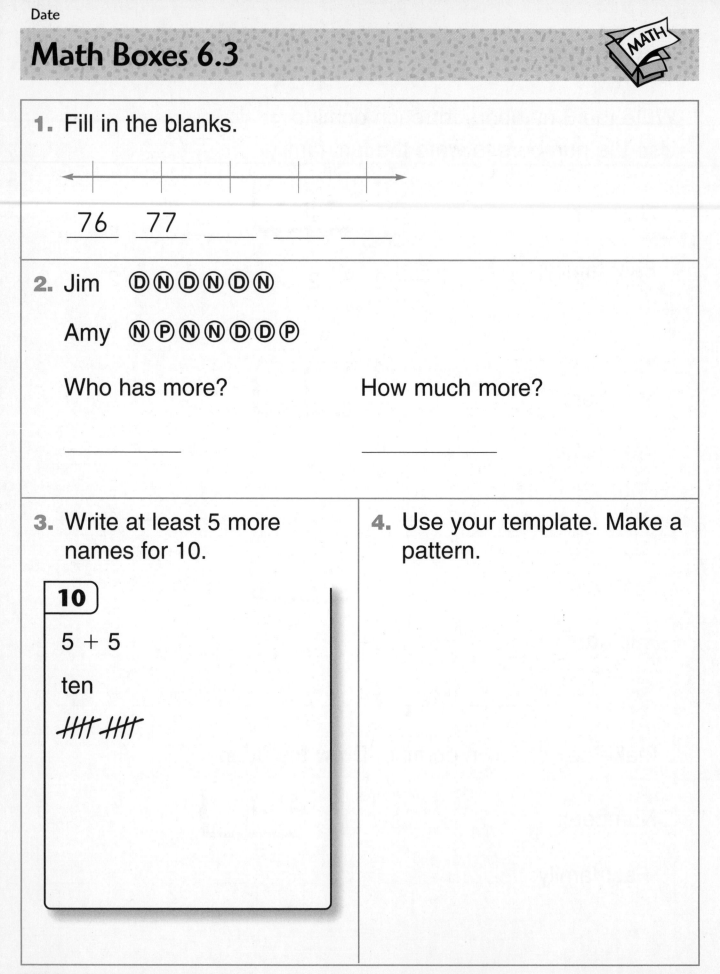

76   77   ___   ___   ___

2. Jim   Ⓓ Ⓝ Ⓓ Ⓝ Ⓓ Ⓝ

Amy   Ⓝ Ⓟ Ⓝ Ⓝ Ⓓ Ⓓ Ⓟ

Who has more?   How much more?

_____   _____

3. Write at least 5 more names for 10.

10
5 + 5
ten
‖‖‖ ‖‖‖

4. Use your template. Make a pattern.

Fact Power Table

0 + 0	0 + 1	0 + 2	0 + 3	0 + 4	0 + 5	0 + 6	0 + 7	0 + 8	0 + 9
1 + 0	1 + 1	1 + 2	1 + 3	1 + 4	1 + 5	1 + 6	1 + 7	1 + 8	1 + 9
2 + 0	2 + 1	2 + 2	2 + 3	2 + 4	2 + 5	2 + 6	2 + 7	2 + 8	2 + 9
3 + 0	3 + 1	3 + 2	3 + 3	3 + 4	3 + 5	3 + 6	3 + 7	3 + 8	3 + 9
4 + 0	4 + 1	4 + 2	4 + 3	4 + 4	4 + 5	4 + 6	4 + 7	4 + 8	4 + 9
5 + 0	5 + 1	5 + 2	5 + 3	5 + 4	5 + 5	5 + 6	5 + 7	5 + 8	5 + 9
6 + 0	6 + 1	6 + 2	6 + 3	6 + 4	6 + 5	6 + 6	6 + 7	6 + 8	6 + 9
7 + 0	7 + 1	7 + 2	7 + 3	7 + 4	7 + 5	7 + 6	7 + 7	7 + 8	7 + 9
8 + 0	8 + 1	8 + 2	8 + 3	8 + 4	8 + 5	8 + 6	8 + 7	8 + 8	8 + 9
9 + 0	9 + 1	9 + 2	9 + 3	9 + 4	9 + 5	9 + 6	9 + 7	9 + 8	9 + 9

(one hundred thirty-nine) **139**

Fact Power Game Mat

$\begin{array}{r}0\\+0\end{array}$	$\begin{array}{r}0\\+1\end{array}$	$\begin{array}{r}0\\+2\end{array}$	$\begin{array}{r}0\\+3\end{array}$	$\begin{array}{r}0\\+4\end{array}$	$\begin{array}{r}0\\+5\end{array}$	$\begin{array}{r}0\\+6\end{array}$	$\begin{array}{r}0\\+7\end{array}$	$\begin{array}{r}0\\+8\end{array}$	$\begin{array}{r}0\\+9\end{array}$
$\begin{array}{r}1\\+0\end{array}$	$\begin{array}{r}1\\+1\end{array}$	$\begin{array}{r}1\\+2\end{array}$	$\begin{array}{r}1\\+3\end{array}$	$\begin{array}{r}1\\+4\end{array}$	$\begin{array}{r}1\\+5\end{array}$	$\begin{array}{r}1\\+6\end{array}$	$\begin{array}{r}1\\+7\end{array}$	$\begin{array}{r}1\\+8\end{array}$	$\begin{array}{r}1\\+9\end{array}$
$\begin{array}{r}2\\+0\end{array}$	$\begin{array}{r}2\\+1\end{array}$	$\begin{array}{r}2\\+2\end{array}$	$\begin{array}{r}2\\+3\end{array}$	$\begin{array}{r}2\\+4\end{array}$	$\begin{array}{r}2\\+5\end{array}$	$\begin{array}{r}2\\+6\end{array}$	$\begin{array}{r}2\\+7\end{array}$	$\begin{array}{r}2\\+8\end{array}$	$\begin{array}{r}2\\+9\end{array}$
$\begin{array}{r}3\\+0\end{array}$	$\begin{array}{r}3\\+1\end{array}$	$\begin{array}{r}3\\+2\end{array}$	$\begin{array}{r}3\\+3\end{array}$	$\begin{array}{r}3\\+4\end{array}$	$\begin{array}{r}3\\+5\end{array}$	$\begin{array}{r}3\\+6\end{array}$	$\begin{array}{r}3\\+7\end{array}$	$\begin{array}{r}3\\+8\end{array}$	$\begin{array}{r}3\\+9\end{array}$
$\begin{array}{r}4\\+0\end{array}$	$\begin{array}{r}4\\+1\end{array}$	$\begin{array}{r}4\\+2\end{array}$	$\begin{array}{r}4\\+3\end{array}$	$\begin{array}{r}4\\+4\end{array}$	$\begin{array}{r}4\\+5\end{array}$	$\begin{array}{r}4\\+6\end{array}$	$\begin{array}{r}4\\+7\end{array}$	$\begin{array}{r}4\\+8\end{array}$	$\begin{array}{r}4\\+9\end{array}$
$\begin{array}{r}5\\+0\end{array}$	$\begin{array}{r}5\\+1\end{array}$	$\begin{array}{r}5\\+2\end{array}$	$\begin{array}{r}5\\+3\end{array}$	$\begin{array}{r}5\\+4\end{array}$	$\begin{array}{r}5\\+5\end{array}$	$\begin{array}{r}5\\+6\end{array}$	$\begin{array}{r}5\\+7\end{array}$	$\begin{array}{r}5\\+8\end{array}$	$\begin{array}{r}5\\+9\end{array}$
$\begin{array}{r}6\\+0\end{array}$	$\begin{array}{r}6\\+1\end{array}$	$\begin{array}{r}6\\+2\end{array}$	$\begin{array}{r}6\\+3\end{array}$	$\begin{array}{r}6\\+4\end{array}$	$\begin{array}{r}6\\+5\end{array}$	$\begin{array}{r}6\\+6\end{array}$	$\begin{array}{r}6\\+7\end{array}$	$\begin{array}{r}6\\+8\end{array}$	$\begin{array}{r}6\\+9\end{array}$
$\begin{array}{r}7\\+0\end{array}$	$\begin{array}{r}7\\+1\end{array}$	$\begin{array}{r}7\\+2\end{array}$	$\begin{array}{r}7\\+3\end{array}$	$\begin{array}{r}7\\+4\end{array}$	$\begin{array}{r}7\\+5\end{array}$	$\begin{array}{r}7\\+6\end{array}$	$\begin{array}{r}7\\+7\end{array}$	$\begin{array}{r}7\\+8\end{array}$	$\begin{array}{r}7\\+9\end{array}$
$\begin{array}{r}8\\+0\end{array}$	$\begin{array}{r}8\\+1\end{array}$	$\begin{array}{r}8\\+2\end{array}$	$\begin{array}{r}8\\+3\end{array}$	$\begin{array}{r}8\\+4\end{array}$	$\begin{array}{r}8\\+5\end{array}$	$\begin{array}{r}8\\+6\end{array}$	$\begin{array}{r}8\\+7\end{array}$	$\begin{array}{r}8\\+8\end{array}$	$\begin{array}{r}8\\+9\end{array}$
$\begin{array}{r}9\\+0\end{array}$	$\begin{array}{r}9\\+1\end{array}$	$\begin{array}{r}9\\+2\end{array}$	$\begin{array}{r}9\\+3\end{array}$	$\begin{array}{r}9\\+4\end{array}$	$\begin{array}{r}9\\+5\end{array}$	$\begin{array}{r}9\\+6\end{array}$	$\begin{array}{r}9\\+7\end{array}$	$\begin{array}{r}9\\+8\end{array}$	$\begin{array}{r}9\\+9\end{array}$

Use with Lesson 6.4.

Math Boxes 6.4

1. Use Ⓓ, Ⓝ, and Ⓟ.

Show 71¢ in two ways.

2. Use <, >, or = with these *Top-It* cards.

| 8 | 6 | | 5 | 10 |

3. Draw a line segment about 1 inch shorter than the one below.

_____

4. Fill in the missing numbers.

in
↓

Rule

Add 5

out

in	out
8	
12	
29	
41	
100	

Using the Addition/Subtraction Facts Table 2

+,−	0	1	2	3	4	5	6	7	8	9
0	0	1	2	3	4	5	6	7	8	9
1	1	2	3	4	5	6	7	8	9	10
2	2	3	4	5	6	7	8	9	10	11
3	3	4	5	6	7	8	9	10	11	12
4	4	5	6	7	8	9	10	11	12	13
5	5	6	7	8	9	10	11	12	13	14
6	6	7	8	9	10	11	12	13	14	15
7	7	8	9	10	11	12	13	14	15	16
8	8	9	10	11	12	13	14	15	16	17
9	9	10	11	12	13	14	15	16	17	18

Add or subtract. Use the table to help you.

1. $5 + 6 =$ _____   2. $11 - 5 =$ _____

3. $8 + 4 =$ _____   4. $12 - 4 =$ _____

5. $10 - 7 =$ _____   6. $15 - 8 =$ _____

7. $13 - 6 =$ _____   8. $18 - 9 =$ _____

9. $14 - 7 =$ _____   10. $16 - 9 =$ _____

Use with Lesson 6.5.

Math Boxes 6.5

1. Draw the missing dots.

14
::::

9
::::

2. Match.

Breakfast -------- P.M.

Dinner ------- A.M.

Bedtime   A.M.

Wake up   P.M.

3. Use your number grid.
Start at 67. Count back 19.

You end up at _____.

_____ = 67 − 19

4. Cross out the wrong names.

12

7 + 5   11 + 2   12 − 0

14 − 3   9 + 3   16 − 5

4 + 5   10 − 2   3 + 9

2 + 11   17 − 5

Measuring in Centimeters

1. Use 2 longs to measure objects in centimeters.
   Record their measures in the table.

Object (Name it or draw it)	My measurement
	about _____ cm
	about _____ cm
	about _____ cm

Use a ruler to measure to the nearest centimeter.

2.

_____

about ____ cm

3. _____

about ____ cm

5. _____

about ____ cm

4.

about ____ cm

6. _____

about ____ cm

7. Draw a line segment 9 centimeters long.

Use with Lesson 6.6.

Math Boxes 6.6

1. Record the time.

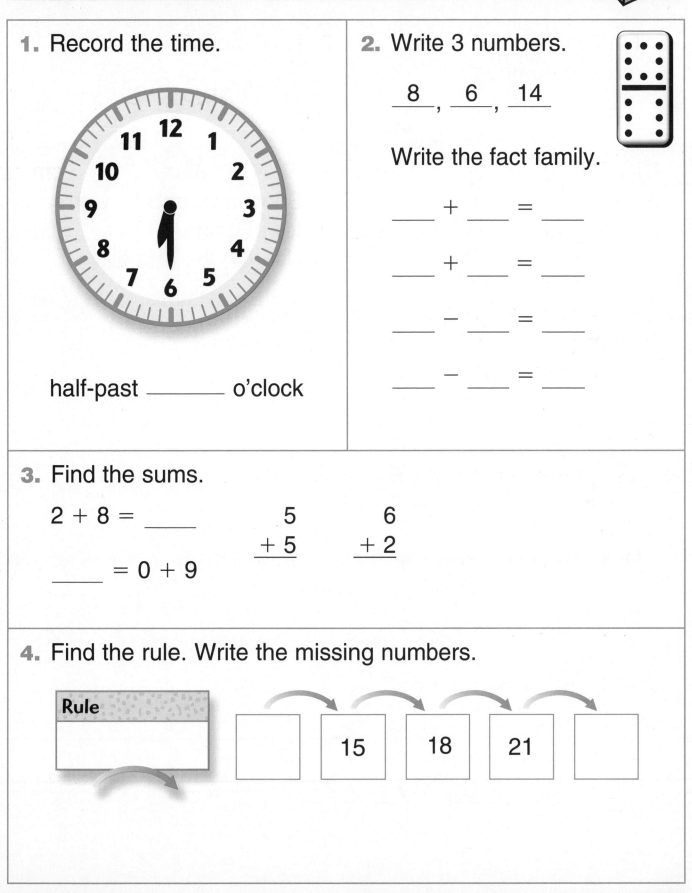

half-past _____ o'clock

2. Write 3 numbers.

____8____, ____6____, ____14____

Write the fact family.

___ + ___ = ___

___ + ___ = ___

___ − ___ = ___

___ − ___ = ___

3. Find the sums.

2 + 8 = ____

____ = 0 + 9

$$\begin{array}{r} 5 \\ + 5 \\ \hline \end{array}$$

$$\begin{array}{r} 6 \\ + 2 \\ \hline \end{array}$$

4. Find the rule. Write the missing numbers.

| Rule | | | 15 | 18 | 21 | |

Math Boxes 6.7

MATH

1. Find the rule. Fill in the missing numbers.

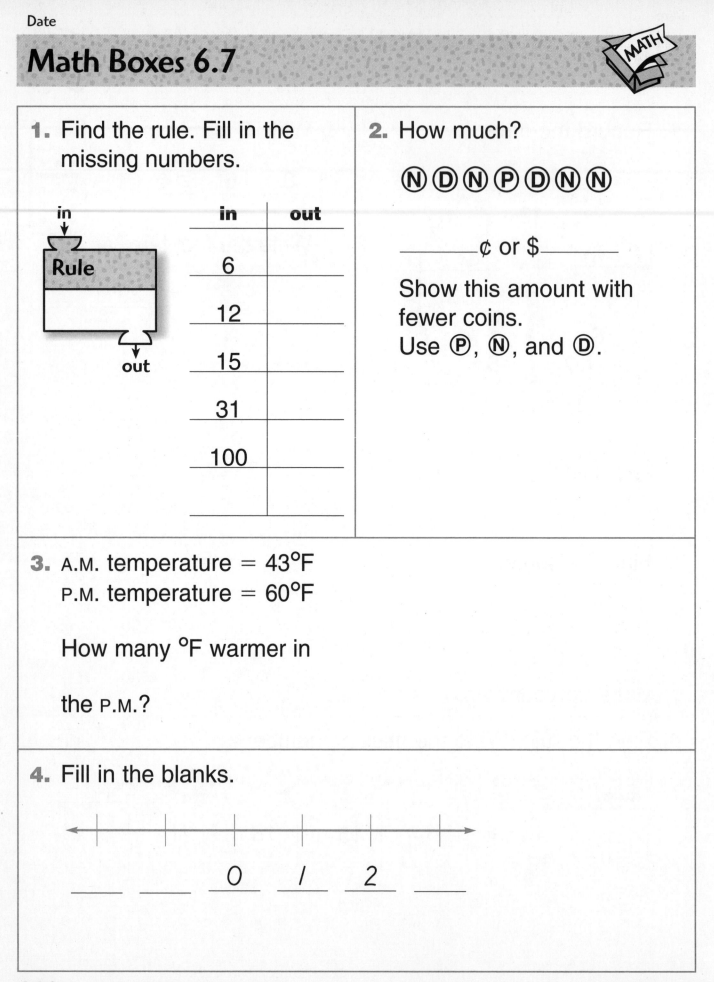

in ↓

Rule

out ↓

in	out
6	
12	
15	
31	
100	

2. How much?

Ⓝ Ⓓ Ⓝ Ⓟ Ⓓ Ⓝ Ⓝ

_____¢ or $_____

Show this amount with fewer coins.
Use Ⓟ, Ⓝ, and Ⓓ.

3. A.M. temperature = 43°F
P.M. temperature = 60°F

How many °F warmer in

the P.M.?

4. Fill in the blanks.

___ ___ ___ 0 / 2 ___

"What's My Rule?"

1. Find the rule.

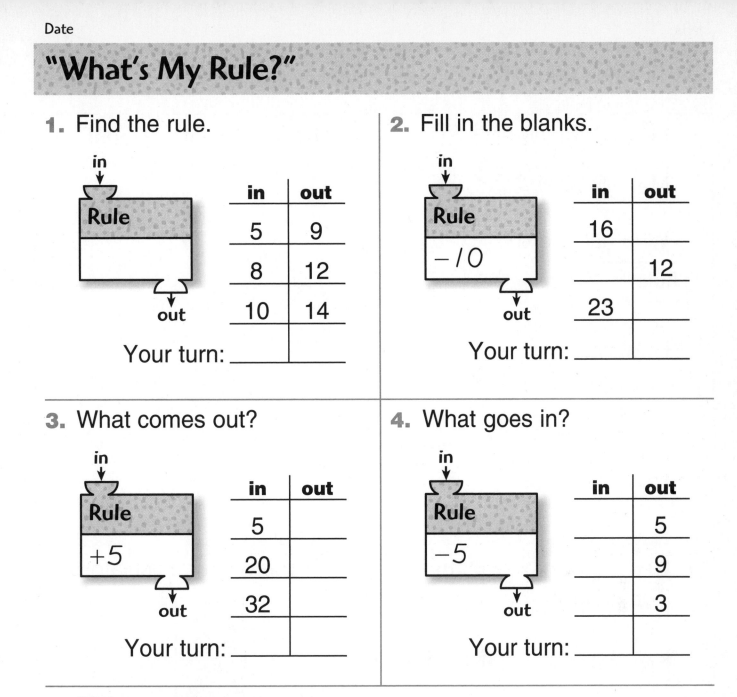

in	out
5	9
8	12
10	14

Your turn: _____

2. Fill in the blanks.

Rule: − 10

in	out
16	
	12
23	

Your turn: _____

3. What comes out?

Rule: +5

in	out
5	
20	
32	

Your turn: _____

4. What goes in?

Rule: −5

in	out
	5
	9
	3

Your turn: _____

5. Make up your own.

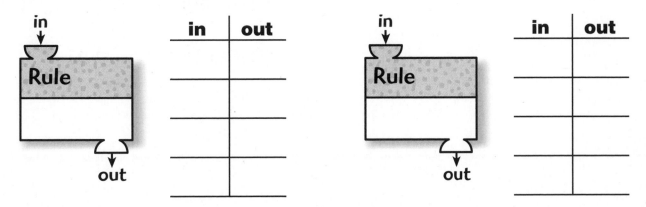

in	out

in	out

Math Boxes 6.8

---

1. Lisa Ⓓ Ⓝ Ⓝ Ⓝ Ⓓ Ⓟ Ⓝ

   Mari Ⓝ Ⓝ Ⓟ Ⓓ Ⓓ Ⓓ Ⓟ

   Who has more?

   _____

   How much more? _____

---

2. Use a calculator. Count up by 25s.

   __0__ , __25__ , __50__ ,

   _____ , _____ , _____ ,

   _____ , _____ , _____ ,

   _____ , _____

---

3. Use <, >, or = with these *Top-It* cards.

   3 **9** ☐ **5** **5**

---

4. Measure your calculator.

   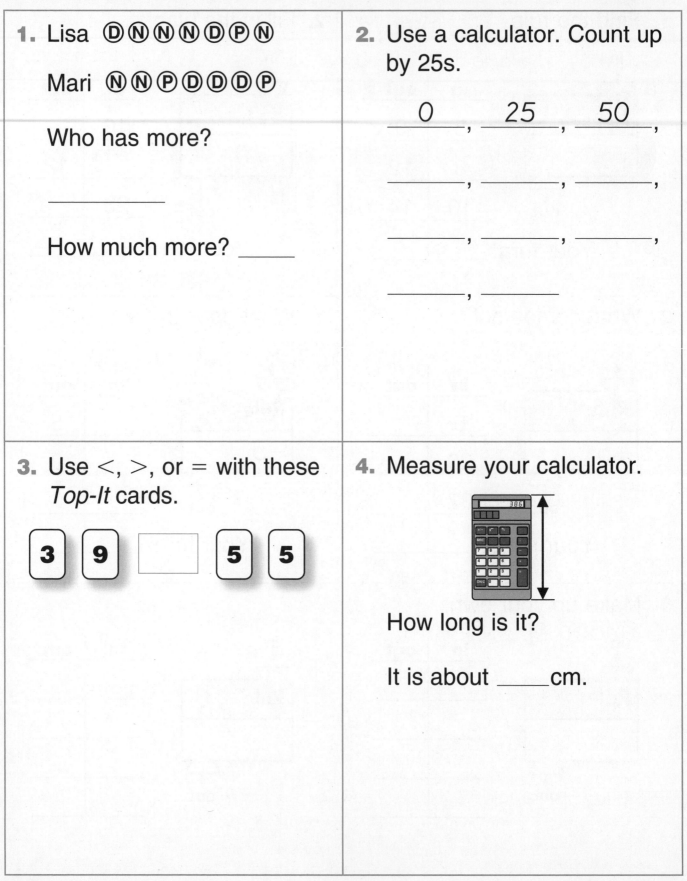

   How long is it?

   It is about _____ cm.

Counting Coins

ⓟ 1¢	Ⓝ 5¢	Ⓓ 10¢	Ⓠ 25¢
$0.01	$0.05	$0.10	$0.25
a penny	a nickel	a dime	a quarter

How much money? Use your coins.

1. _____ ¢ or $_____

2. _____ ¢ or $_____

3. Ⓠ Ⓠ Ⓓ Ⓝ Ⓝ Ⓝ Ⓝ   _____ ¢ or $_____

4. ⓟ Ⓓ Ⓠ ⓟ Ⓓ Ⓠ Ⓠ   _____ ¢ or $_____

5. Ⓠ Ⓠ Ⓓ Ⓠ Ⓠ Ⓝ Ⓠ   _____ ¢ or $_____

6. Make up your own.

_____ ¢ or $_____

1- and 10-Centimeter Objects

Find 3 things that are about 1 centimeter long.

Use words or pictures to show the things you found.

Find 3 things that are about 10 centimeters long.

Use words or pictures to show the things you found.

1. Record the time.

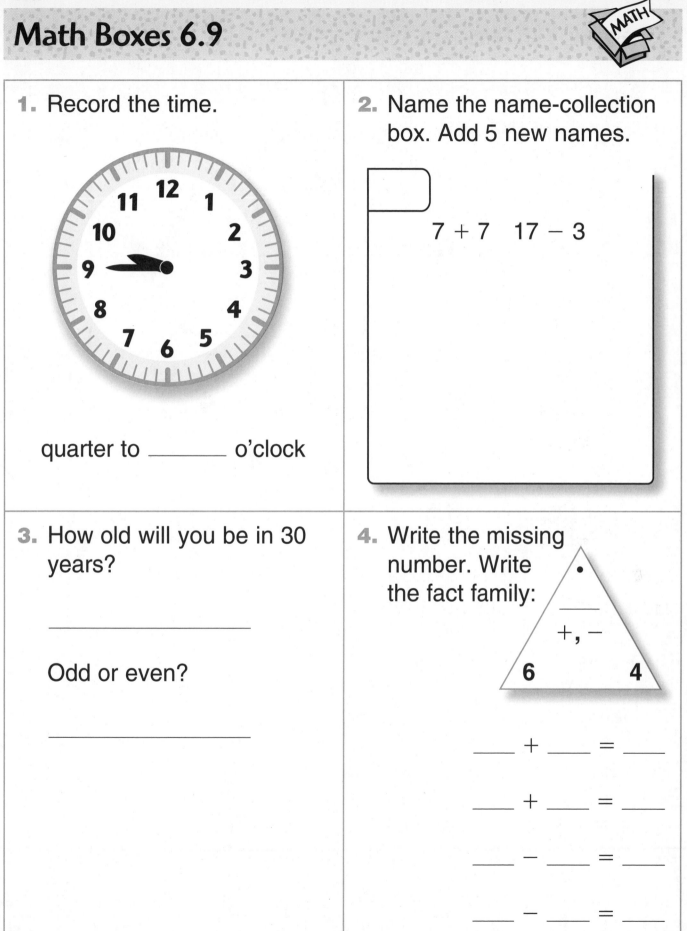

quarter to _____ o'clock

2. Name the name-collection box. Add 5 new names.

7 + 7   17 − 3

3. How old will you be in 30 years?

_____

Odd or even?

_____

4. Write the missing number. Write the fact family:

_____
+, −

6   4

____ + ____ = ____

____ + ____ = ____

____ − ____ = ____

____ − ____ = ____

Time at 5-Minute Intervals

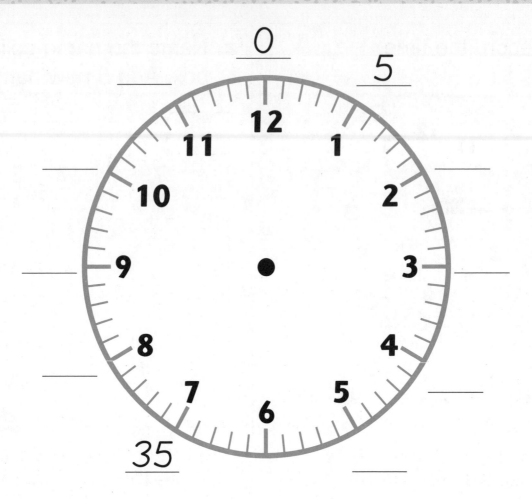

How many minutes are there in:

1. 1 hour? _____ minutes

2. Half an hour? _____ minutes

3. A quarter hour? _____ minutes

4. Three-quarters of an hour? _____ minutes

Digital Notation

Draw the hour hand and the minute hand.

1. 4:00

2. 2:30

3. 6:15

Write the time.

4. _____:_____

5. _____:_____

6. _____:_____

Make up your own. Draw the hour hand and minute hand. Write the time.

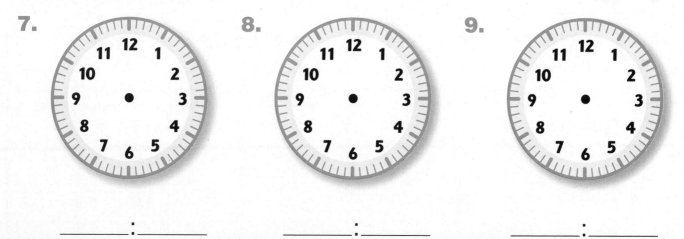

7. _____:_____

8. _____:_____

9. _____:_____

1. How much?

Q N D N D P N

_____¢ or $_____

Use P, N, D, and Q to show this amount with fewer coins.

2. Use your number grid. Start at 48. Count back 15.

You end up at _____.

_____ = 48 − 15

3. Fill in the missing numbers.

in
↓

Rule

+3

out
↓

in	out
	5
	13
	20
	28

4. Measure your shoe.

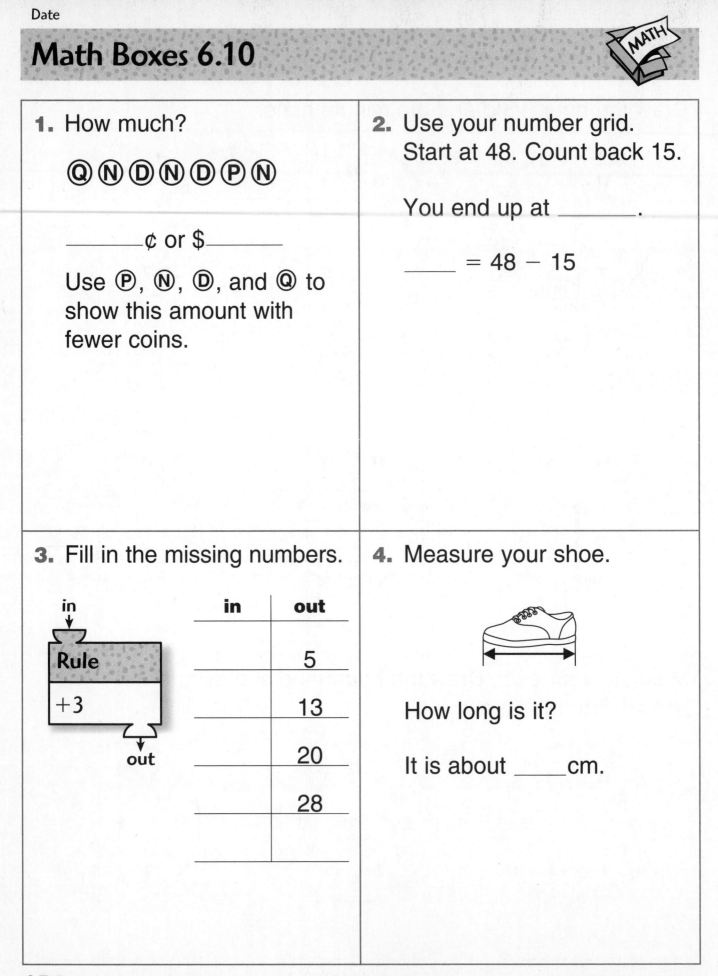

How long is it?

It is about ____cm.

Timing Me

1. I can count to _____ in 15 seconds.

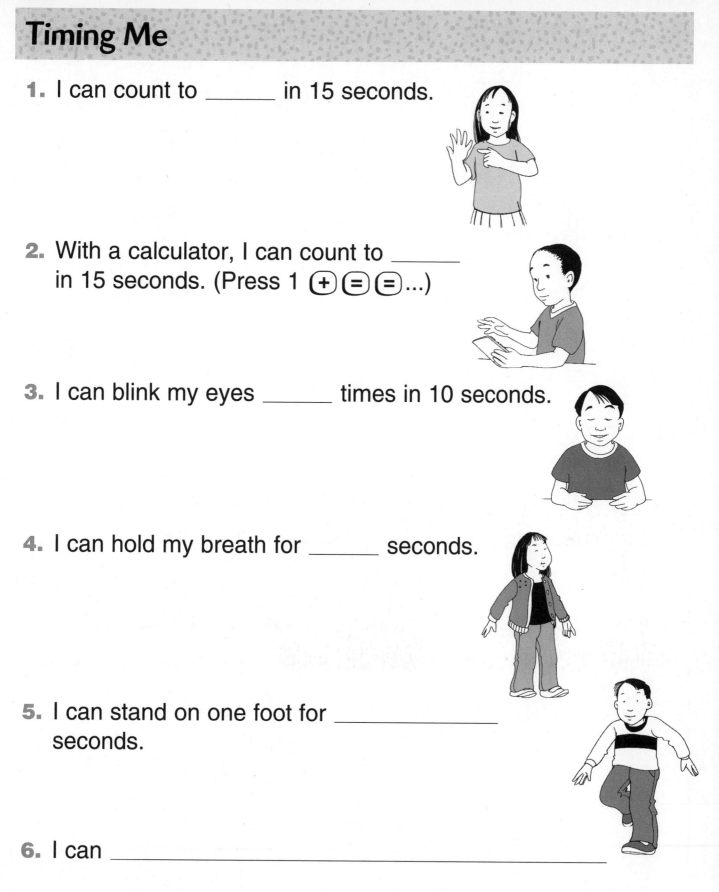

2. With a calculator, I can count to _____ in 15 seconds. (Press 1 (+) (=) (=) ...)

3. I can blink my eyes _____ times in 10 seconds.

4. I can hold my breath for _____ seconds.

5. I can stand on one foot for _____ seconds.

6. I can _____

_____

Math Boxes 6.11

1. What day of the week is today?

_____

What day of the month?

_____

What day of school?

_____

2. Use Ⓠ, Ⓓ, Ⓝ, and Ⓟ.

Show 78¢ in two ways.

3. Find the sums.

3 + 0 = _____

_____ = 9 + 9

$$\begin{array}{r} 7 \\ + 1 \\ \hline \end{array} \qquad \begin{array}{r} 1 \\ + 9 \\ \hline \end{array}$$

4. Draw the hands.

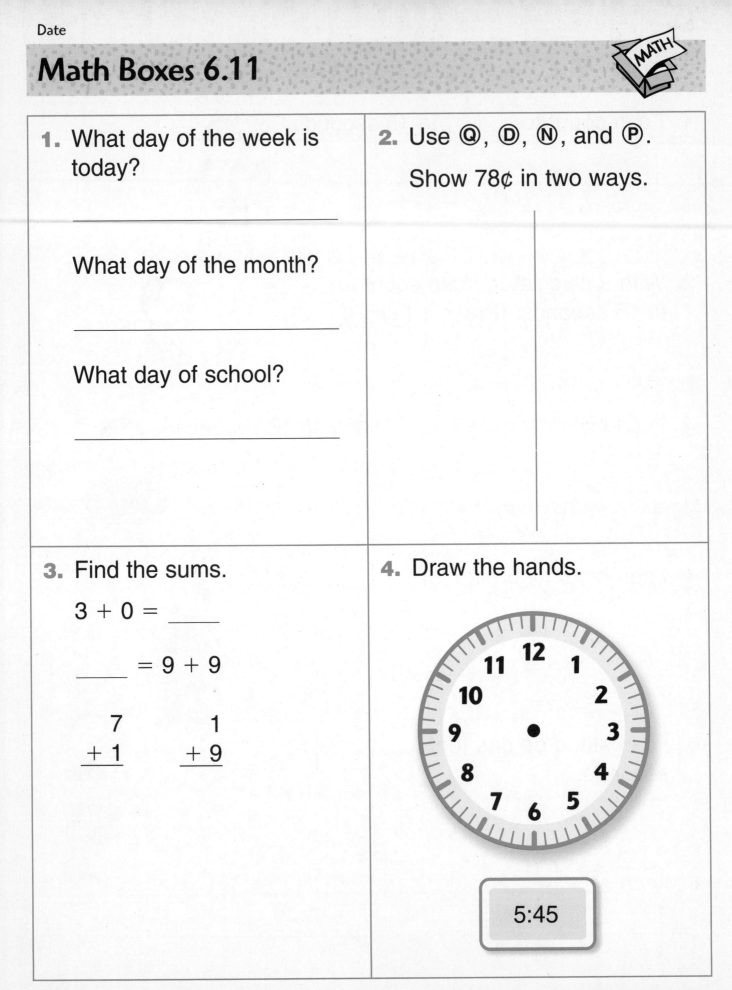

5:45

Use with Lesson 6.11.

Class Results of Calculator Counts

1. I counted to _____ in 15 seconds.

2. Class results:

Largest count	Smallest count	Range of class counts	Middle value of class counts
_____	_____	_____	_____

3. Make a bar graph of the results.

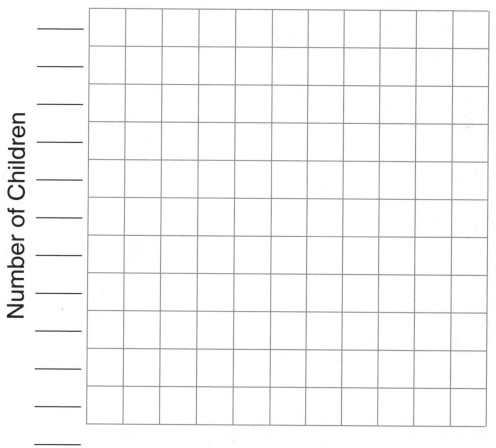

Results of Calculator Counts

Number of Children

Counted to

Math Boxes 6.12

1. Record the time.

:

2. Measure to the nearest centimeter.

_____

_____cm

_____

_____cm

3. Fred ⒟⒬⒟Ⓝ

Jewel ⒟⒟Ⓝ⒟⒟Ⓟ

Who has more?

_____

How much more?

_____

4. Label the box.
Add 5 names.

‖‖‖‖ ‖‖‖‖ ‖‖‖‖   20 − 5

7 + 8

Use with Lesson 6.12.

Math Boxes 6.13

1. Write the numbers.

What am I? ____

What am I? ____

2. Find the rule. Fill in the missing numbers.

in
↓

Rule

out
↓

in	out
15	5
21	11
84	74
30	
	94

3. Write <, >, or =.

Ⓠ Ⓠ ☐ $0.25

$1.00 ☐ Ⓠ Ⓓ Ⓠ Ⓓ

10¢ + 20¢ ☐ Ⓝ Ⓝ Ⓠ

50¢ + 15¢ ☐ Ⓠ Ⓠ Ⓠ

4. Write 3 numbers.

_5_ , _9_ , _14_

Write the fact family.

___ + ___ = ___

___ + ___ = ___

___ − ___ = ___

___ − ___ = ___

Coin Exchanges

Show each amount with fewer coins. Write Ⓟ for penny,
Ⓝ for nickel, Ⓓ for dime, Ⓠ for quarter.

1. _____

    _____

2. _____

    _____

3. _____

    _____

4. Ⓟ Ⓝ Ⓝ Ⓝ Ⓟ Ⓓ Ⓓ Ⓓ _____

5. Ⓝ Ⓓ Ⓝ Ⓝ Ⓝ Ⓟ Ⓝ Ⓠ Ⓓ Ⓝ _____

6. Ⓓ Ⓓ Ⓓ Ⓟ Ⓓ Ⓝ Ⓟ Ⓠ Ⓓ Ⓓ Ⓝ Ⓓ _____

Date

Math Boxes 7.1

1. Complete the table.

Before	Number	After
81	82	83
	39	
	58	
	100	
	147	

2. Write four odd numbers with 4 in the tens place.

_____

_____

_____

_____

3. I bought a ring for $0.18. I gave a @. How much change should I get back?

_____

4. Draw the hands.

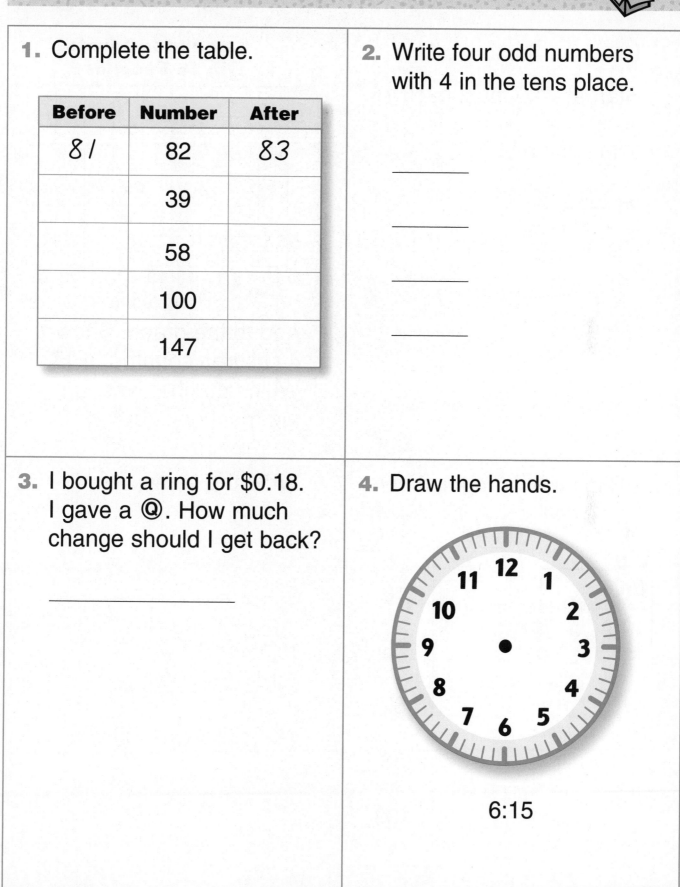

6:15

Math Boxes 7.2

1. Use your number grid.

Start at 71. Count back 14.

You end up at _____.

$71 - 14 =$ _____

2.

Calculator Counts in 15 Seconds

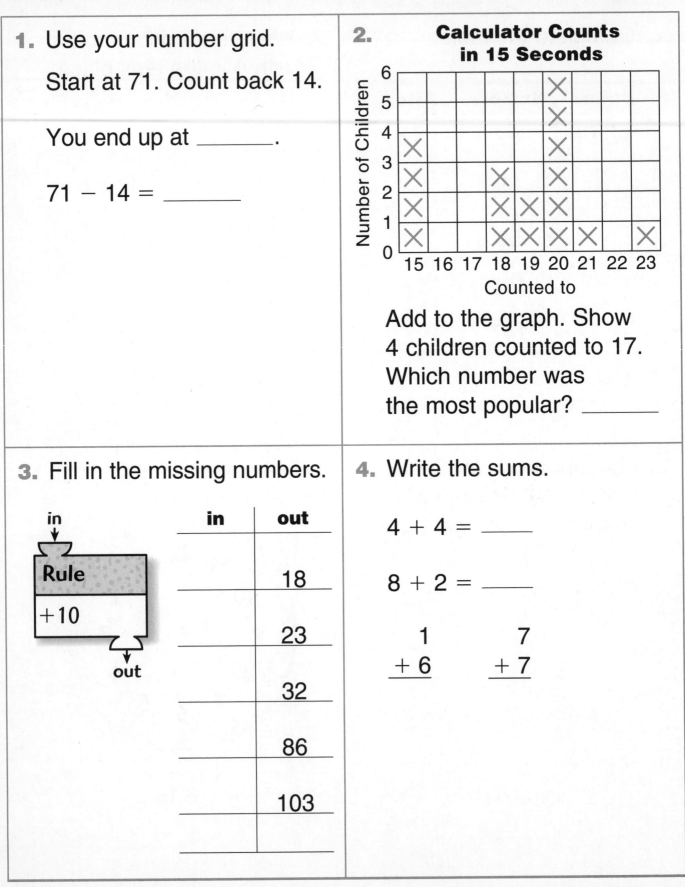

Add to the graph. Show
4 children counted to 17.
Which number was
the most popular? _____

3. Fill in the missing numbers.

in

Rule
+10

out

in	out
	18
	23
	32
	86
	103

4. Write the sums.

$4 + 4 =$ _____

$8 + 2 =$ _____

$$\begin{array}{r} 1 \\ + 6 \\ \hline \end{array} \qquad \begin{array}{r} 7 \\ + 7 \\ \hline \end{array}$$

Use with Lesson 7.2.

1. Write the fact family.

___ + ___ = ___

___ + ___ = ___

___ − ___ = ___

___ − ___ = ___

2. Fill in the label.
Add 5 names.

18 − 10   2 + 6

3. Count up by 1s.

_262_ , _263_ , _264_ ,

_____ , _____ , _____ ,

_____ , _____ , _____ ,

_____ , _____

4. Jonah Ⓠ Ⓓ Ⓓ Ⓠ Ⓟ Ⓟ Ⓟ

_____¢

Mari Ⓠ Ⓓ Ⓝ Ⓝ Ⓓ Ⓠ Ⓓ

_____¢

Who has more? _____

How much more? _____

Date

Pattern-Block Template Shapes

1. Use your template to draw each shape.

square	large triangle	small hexagon
trapezoid	small triangle	fat rhombus
large circle	skinny rhombus	large hexagon

Use with Lesson 7.3.

Pattern-Block Template Shapes (cont.)

2. Draw the shapes that have exactly 4 sides and
   4 corners. Write their names.

_____   _____

_____   _____

Date _____

Line Segments

1. Measure the line segments below to the nearest inch.

   _____   _____ in.

   _____   _____ in.

   _____   _____ in.

2. Draw a line segment that is 5 inches long.

3. Measure the line segments below to the nearest centimeter.

   _____   _____ cm

   _____   _____ cm

   _____   _____ cm

4. Draw a line segment that is 5 centimeters long.

5. Draw a line segment that is 15 centimeters long.

Challenge

6. How much longer is your 15-centimeter line segment

   than your 5-centimeter line segment?   _____ cm

Polygons

Triangles

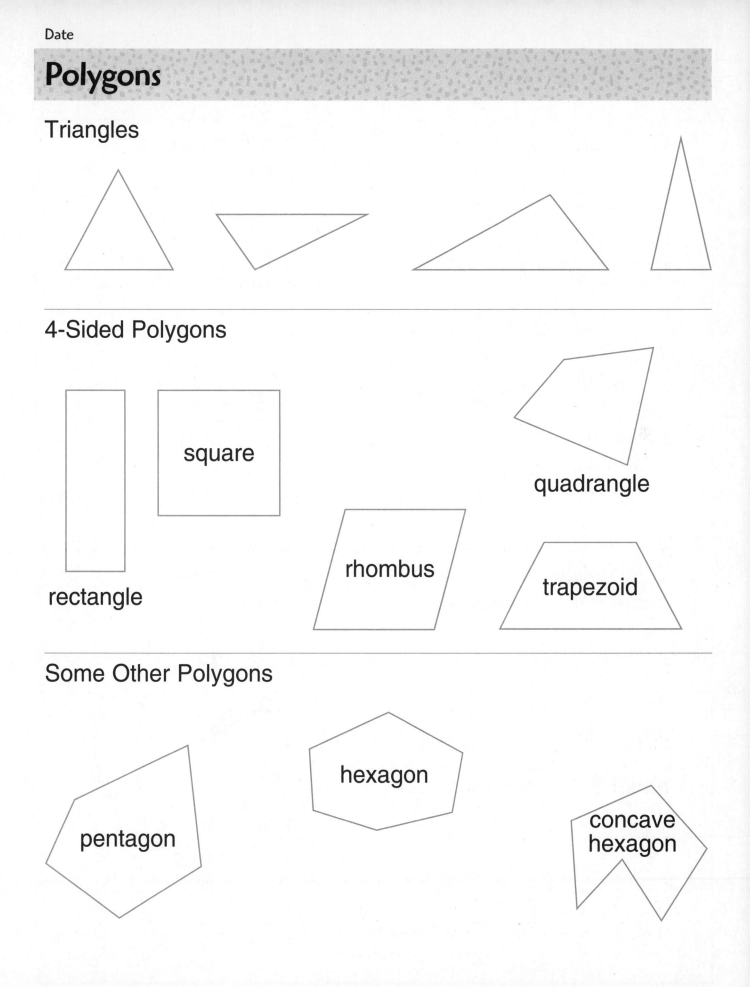

4-Sided Polygons

square

quadrangle

rhombus

trapezoid

rectangle

Some Other Polygons

pentagon

hexagon

concave hexagon

Math Boxes 7.4

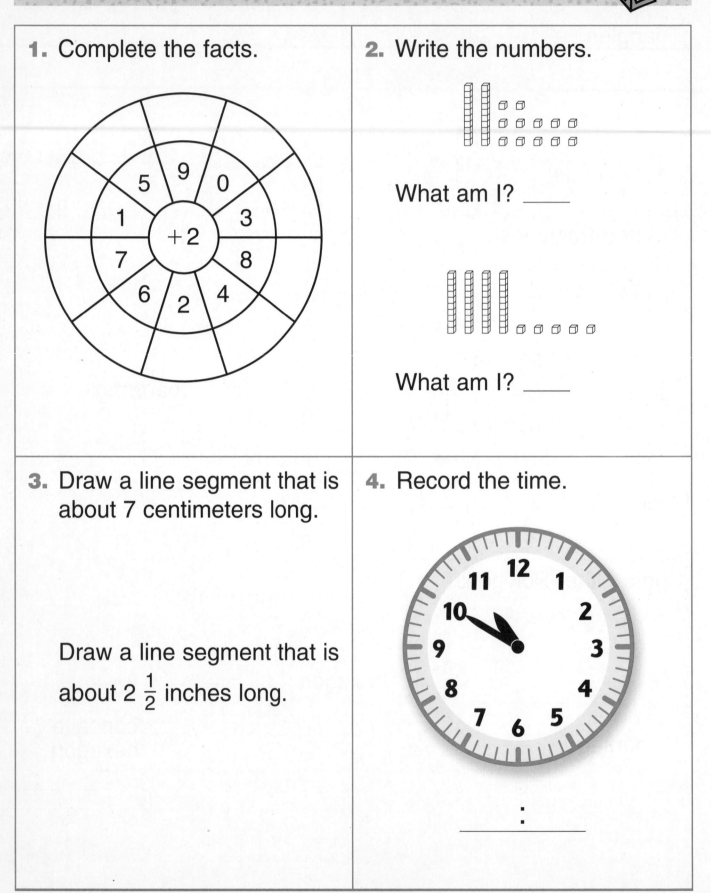

1. Complete the facts.

9
5   0
1   +2   3
7   8
6   2   4

2. Write the numbers.

What am I? ____

What am I? ____

3. Draw a line segment that is about 7 centimeters long.

Draw a line segment that is about $2 \frac{1}{2}$ inches long.

4. Record the time.

____ : ____

Use with Lesson 7.4.

1. Use your template. Make a pattern with small ◯s and △s.

2. Fill in the missing numbers.

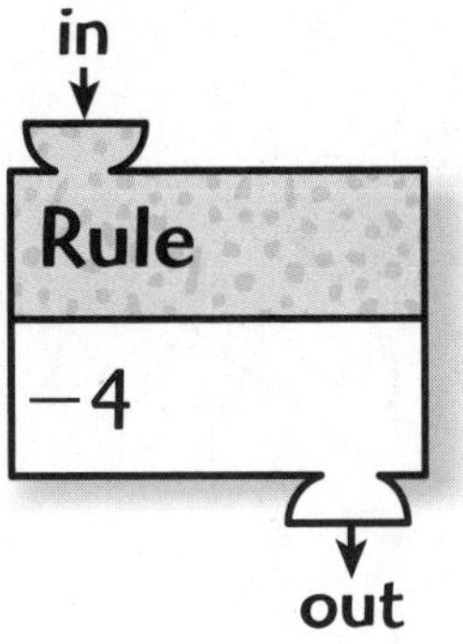

in

Rule
−4

out

in	out
40	36
	9
	23
51	
80	

3. How old will you be in 12 years?

_____

Odd or even?

_____

4. Yes or no?

$0.73 > 91¢ _____

114¢ < $1.03 _____

6 dimes < 80¢ _____

ⓆⓆⓆⓆ = $1.00 _____

Date

3-Dimensional Shapes Poster

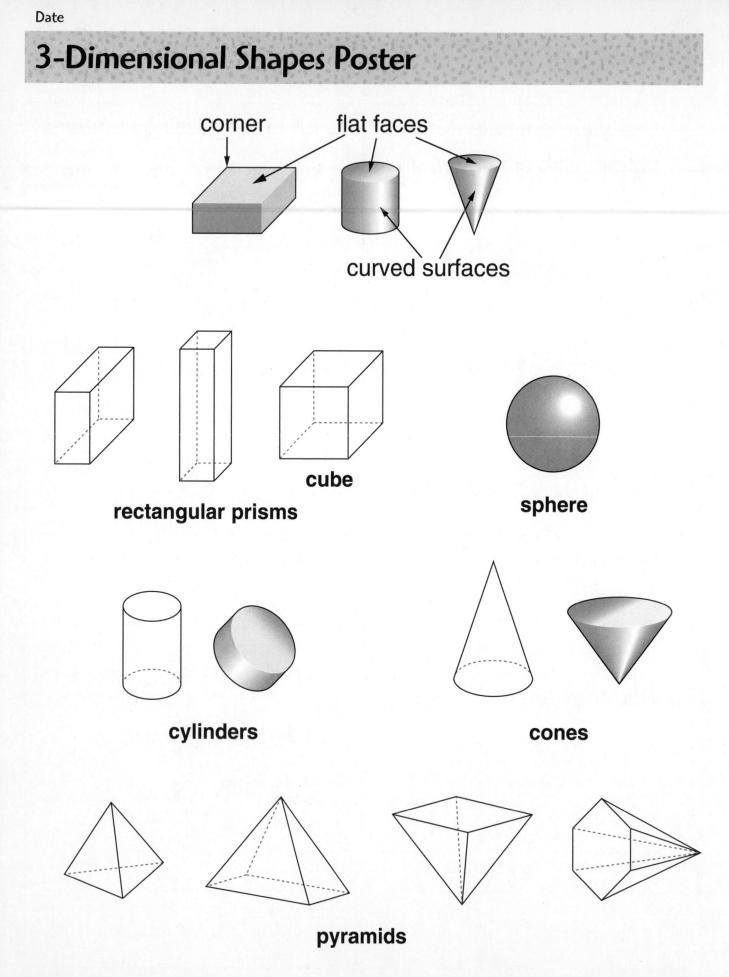

corner

flat faces

curved surfaces

rectangular prisms

cube

sphere

cylinders

cones

pyramids

Use with Lesson 7.6.

Identifying 3-Dimensional Shapes

What kind of a shape is each object? Write its name under the picture.

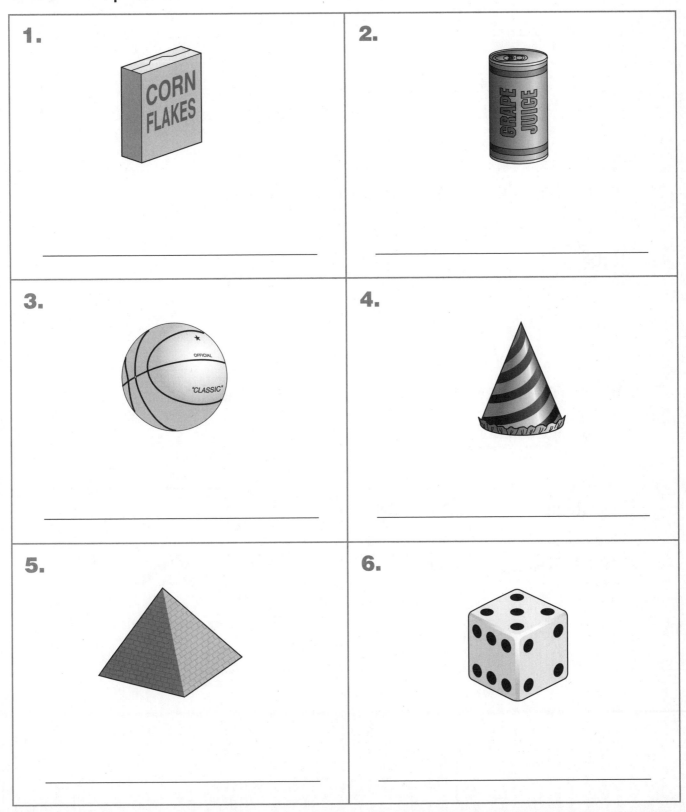

1. CORN FLAKES

2. GRAPE JUICE

3.

4.

5.

6.

Math Boxes 7.6

1. Write the fact family.

___ + ___ = ___   ___ − ___ = ___

___ + ___ = ___   ___ − ___ = ___

10

+, −

7   3

2. Fill in the rule and the missing numbers.

Rule

| | | 148 | 158 | |

3. Measure a crayon.

CRAYON

It is about ___ centimeters.

It is about ___ inches.

4. Tell the time.

___:___

Use with Lesson 7.6.

1. Fill in the oval next to the name of the shape.

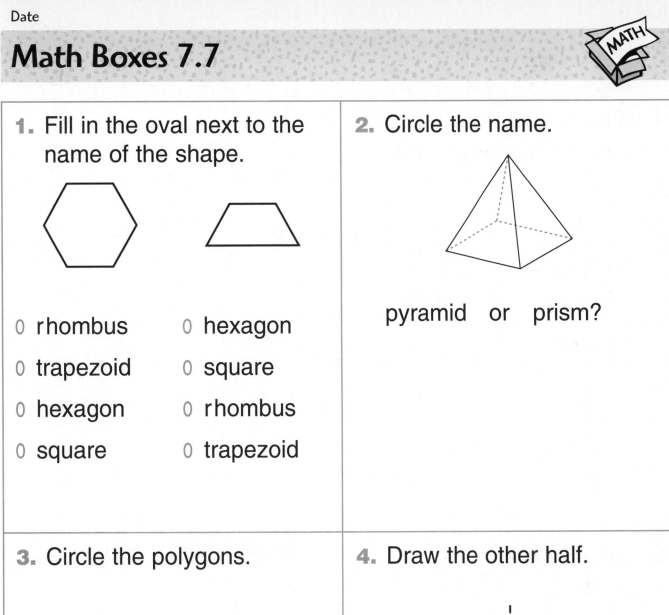

0 rhombus   0 hexagon

0 trapezoid   0 square

0 hexagon   0 rhombus

0 square   0 trapezoid

2. Circle the name.

pyramid   or   prism?

3. Circle the polygons.

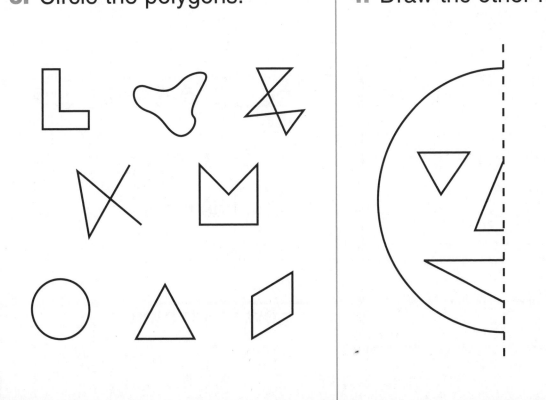

4. Draw the other half.

Math Boxes 7.8

MATH

1. Write two-digit numbers with 7 in the ones place.

____ ____ ____

Write two-digit numbers with 7 in the tens place.

____ ____ ____

2. Write <, >, or =.

3 + 9 ☐ 10

18 ☐ 20 − 10

59 ☐ 79

8 + 8 ☐ 9 + 10

3. Write 5 more names.

| 100 | 80 + 20 |

4. Marcy Ⓠ Ⓓ Ⓝ Ⓟ Ⓝ Ⓝ Ⓓ

_____ ¢

Claudia Ⓓ Ⓓ Ⓝ Ⓠ Ⓝ Ⓠ Ⓟ

_____ ¢

Who has more?

_____

How much more? _____

Date

Record the amount shown.

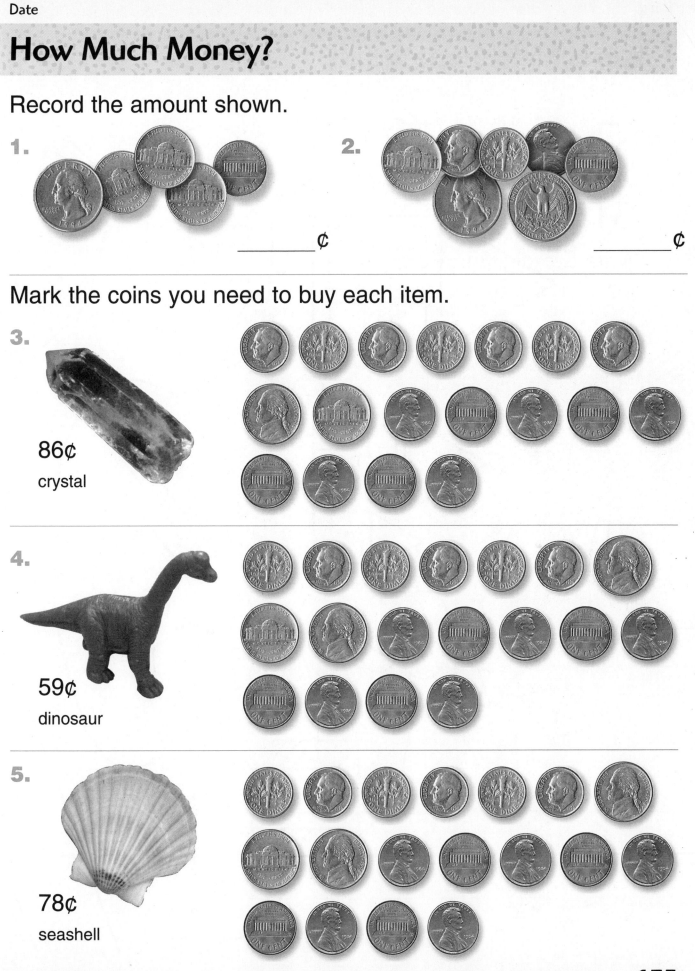

1. _____ ¢

2. _____ ¢

Mark the coins you need to buy each item.

3.

86¢
crystal

4.

59¢
dinosaur

5.

78¢
seashell

Shapes Neighborhood

Use with Lesson 8.1.

Shapes Neighborhood Tally

Tally the shapes you find on page 176.

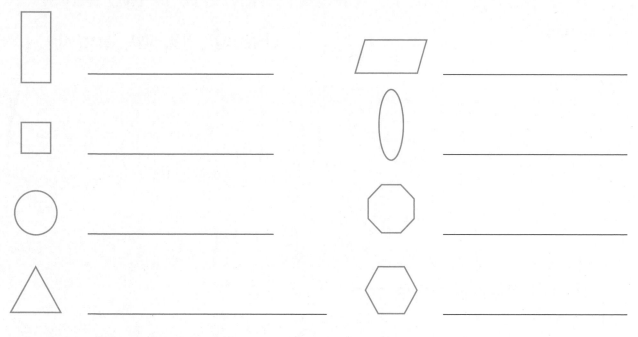

Count the tally marks for each shape. How many?

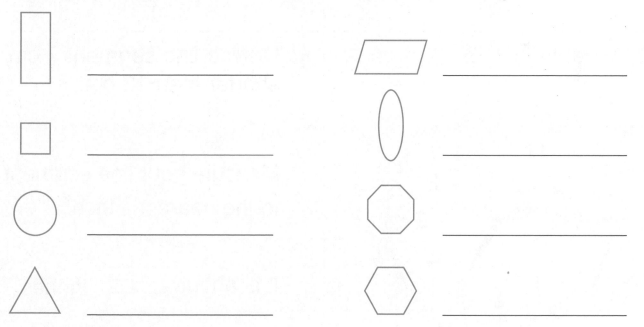

Color the Shapes Neighborhood.

Math Boxes 8.1

1. Draw a polygon with 6 sides.

Is the number of sides odd or even?

_____

2. Show 81¢ in two ways.

Use Ⓠ, Ⓓ, Ⓝ, and Ⓟ.

3. What time is it?

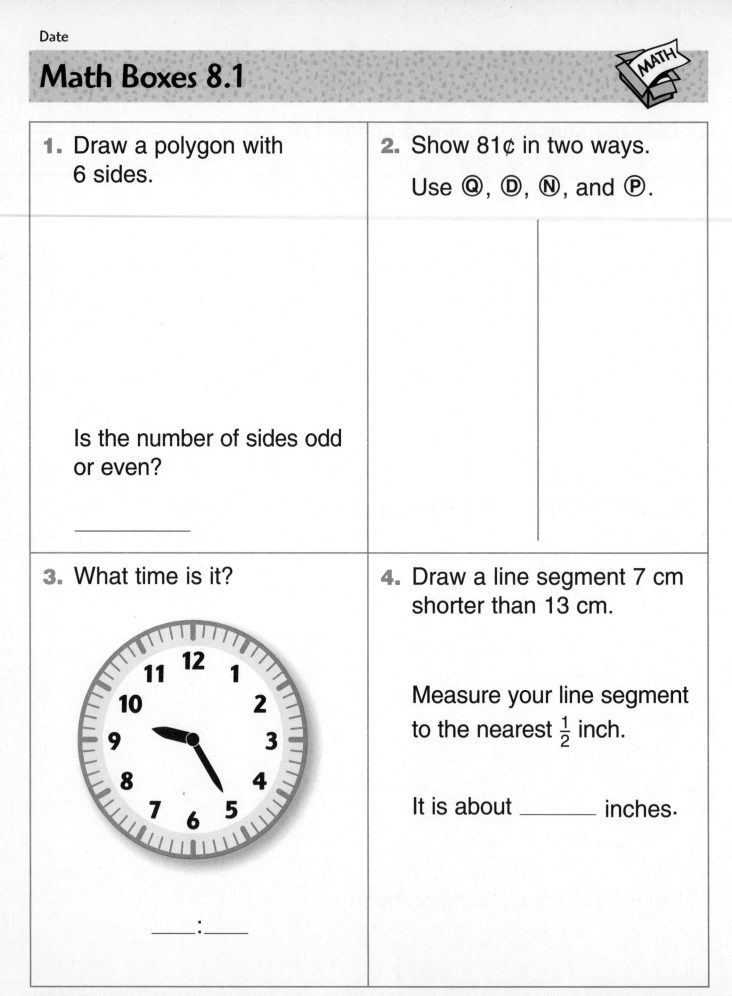

____ : ____

4. Draw a line segment 7 cm shorter than 13 cm.

Measure your line segment to the nearest $\frac{1}{2}$ inch.

It is about _____ inches.

Use with Lesson 8.1.

Place-Value Mat

$0.01
1¢

Pennies 1s Cubes

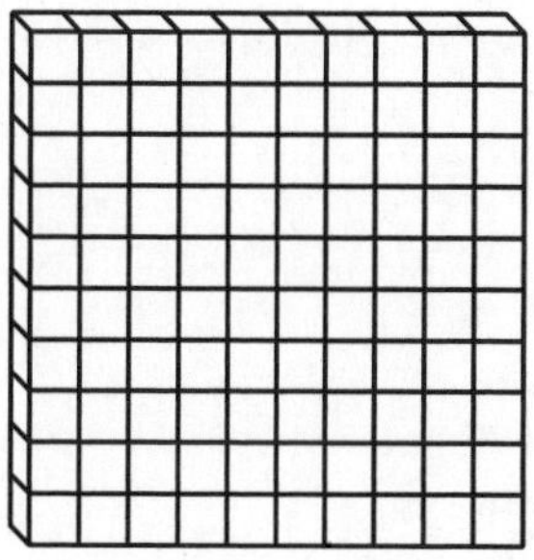

$0.10
10¢

Dimes 10s Longs

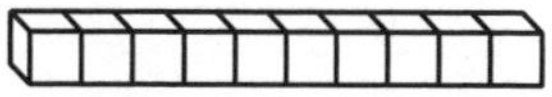

$1.00
100¢

Dollars 100s Flats

Game Report and Review

Game Report

Dollars $1.00	Dimes $0.10	Pennies $0.01	Total
			$_____._____

The Broken Calculator

A key on your calculator is broken. Can you still use your calculator? Show how. Make up your own on the last line.

Broken key	Show in display.	How?
5	15	10 + 2 + 3
8	18	
3	23	
0	20	

Review

Write <, >, or =.

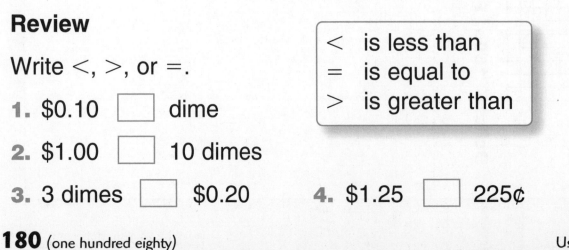

<	is less than
=	is equal to
>	is greater than

1. $0.10 ☐ dime

2. $1.00 ☐ 10 dimes

3. 3 dimes ☐ $0.20   4. $1.25 ☐ 225¢

1. Complete the table.

Before	Number	After
134	135	136
	39	
	61	
	109	
	299	

2. A toy elephant costs $0.72.
I gave 3@s.

How much change will
I get?

_____

3. Write the fact family.

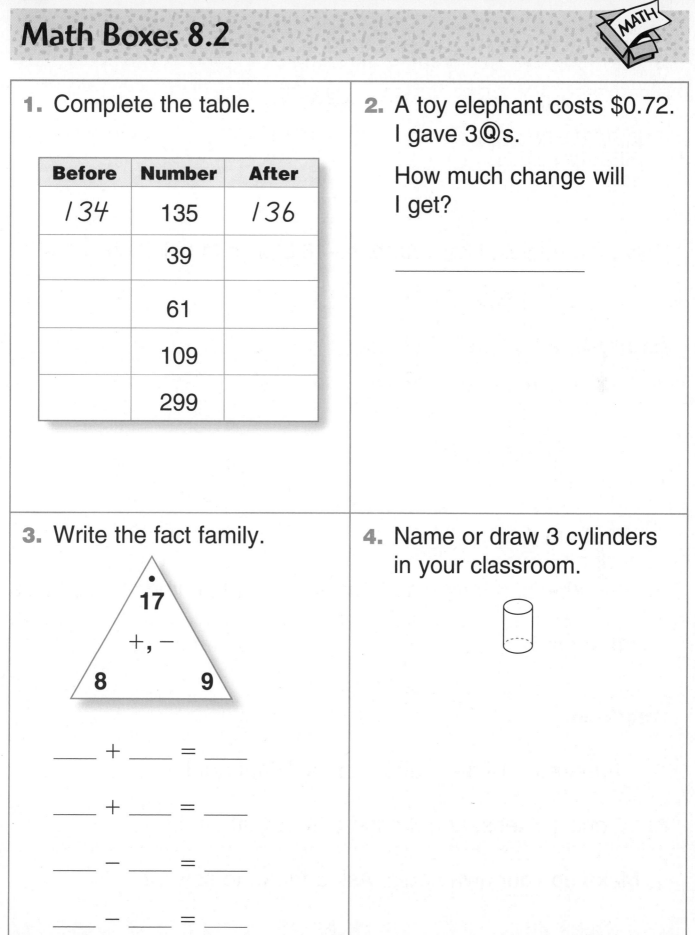

17

+, −

8   9

____ + ____ = ____

____ + ____ = ____

____ − ____ = ____

____ − ____ = ____

4. Name or draw 3 cylinders
in your classroom.

Hundreds, Tens, and Ones Riddles

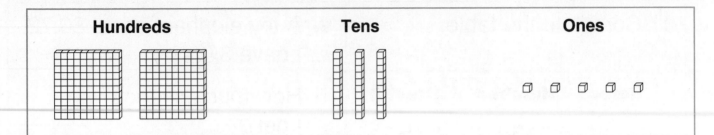

Hundreds	Tens	Ones

Solve the riddles. Use your base-10 blocks to help you.

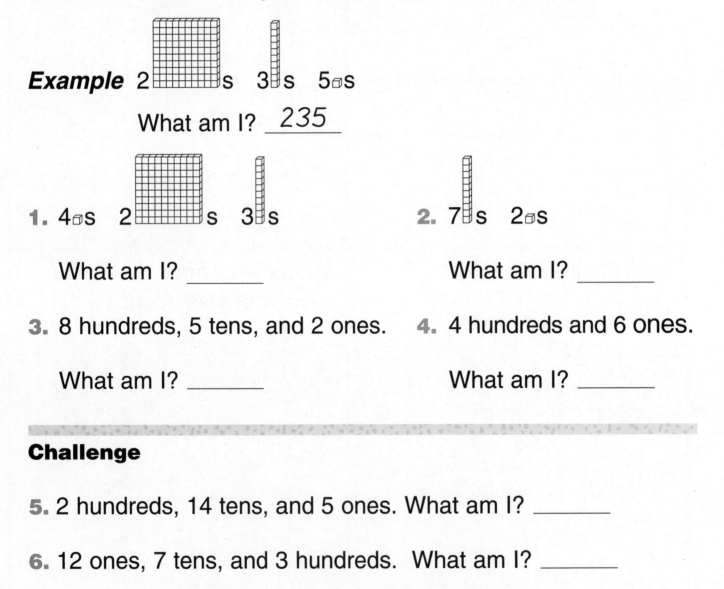

Example 2 ▢s   3 |s   5 ▫s

What am I? __235__

1. 4 ▫s   2 ▢s   3 |s

What am I? _____

2. 7 |s   2 ▫s

What am I? _____

3. 8 hundreds, 5 tens, and 2 ones.

What am I? _____

4. 4 hundreds and 6 ones.

What am I? _____

Challenge

5. 2 hundreds, 14 tens, and 5 ones. What am I? _____

6. 12 ones, 7 tens, and 3 hundreds. What am I? _____

7. Make up your own riddle. Ask a friend to solve it.

Measurement Review

Measure the line segments.

1. |——————|

about _____ cm

2. |———————————————|

about _____ cm

Draw line segments that are these lengths.

3. 15 cm

|- -

4. 8 cm

|- -

5. |————————————————| This line segment is 5 cm long.

Draw a line segment that is 7 cm longer.

|- -

6. Connect the dots in order. Use a straightedge.

Find 3 △s.

Challenge

7. Try to find 4 △s.

8. Color a 4-sided figure.

Math Boxes 8.3

1. Fill in the label. Cross out names that do not belong. Add 2 names.

Ⓠ Ⓠ

40 + 1   51 − 1

30 + 20   49 + 1

Ⓓ Ⓓ Ⓓ Ⓓ Ⓓ Ⓓ

2. Use the digits 5 and 7. Write the smallest and largest numbers.

Smallest _____

Largest _____

3. Use a tape measure.

Measure around someone's wrist.

It is about

_____cm

Odd or even?

_____

4. Write <, >, or =.

10¢ ☐ $0.12

Ⓓ Ⓓ Ⓓ ☐ 40¢

$1.00 ☐ Ⓠ Ⓠ Ⓠ

Ⓠ Ⓠ Ⓓ ☐ Ⓝ Ⓝ Ⓝ Ⓓ Ⓓ Ⓓ Ⓠ

Number Stories

Sample Story

I bought a 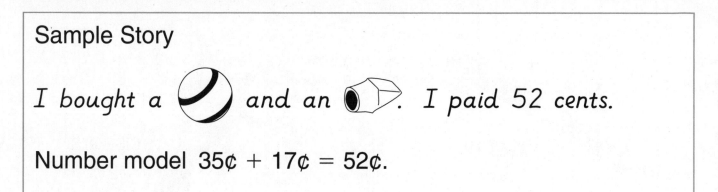 and an ⬤▷. I paid 52 cents.

Number model 35¢ + 17¢ = 52¢.

Story 1

Number model _____

Story 2

Number model _____

crayon
6¢

scissors
32¢

ball
35¢

gum
2¢

pencil
28¢

candy
8¢

eraser
17¢

School Store Mini-Poster 3

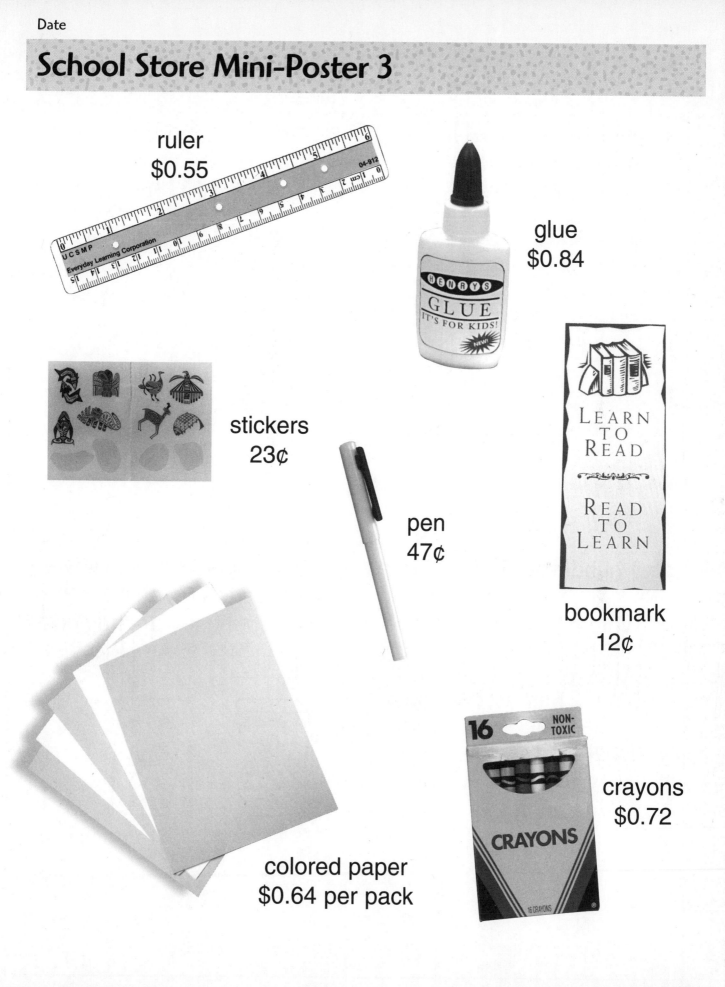

ruler
$0.55

glue
$0.84

stickers
23¢

pen
47¢

bookmark
12¢

colored paper
$0.64 per pack

crayons
$0.72

1. Make points. Use a straightedge. Draw line segments to make a polygon.

2. Draw the hands.

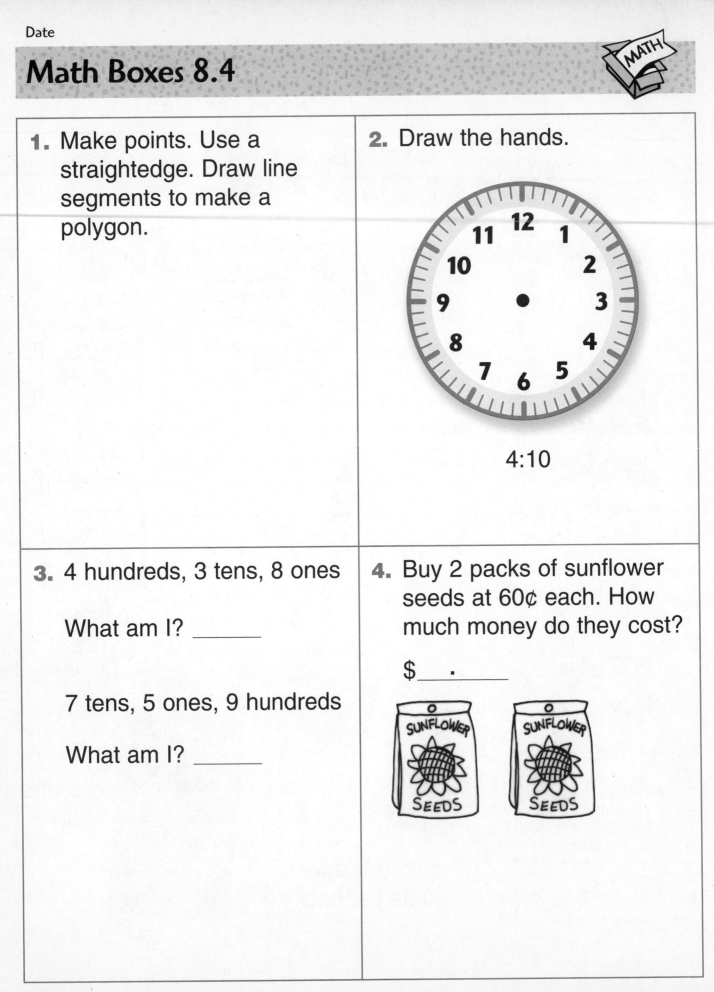

4:10

3. 4 hundreds, 3 tens, 8 ones

What am I? _____

7 tens, 5 ones, 9 hundreds

What am I? _____

4. Buy 2 packs of sunflower seeds at 60¢ each. How much money do they cost?

$_____ . _____

SUNFLOWER SEEDS

SUNFLOWER SEEDS

Museum Store Mini-Poster

seashell
48¢

kite
$1.86

elephant
72¢

rock
35¢

dinosaur
59¢

ring
18¢

magnet
$1.39

puzzle
85¢

plane
27¢

Making Change

Record what you bought. Record how much change you got.

Example

I bought ___a plane___.

The ___plane___ costs ___27___ cents.

I gave ___ⒹⒹⒹ___ to the clerk.

I got ___3___ cents in change.

1. I bought _____.

   The _____ costs _____ cents.

   I gave _____ to the clerk.

   I got _____ cents in change.

2. I bought _____.

   The _____ costs _____ cents.

   I gave _____ to the clerk.

   I got _____ cents in change.

3. I bought _____.

   The _____ costs _____ cents.

   I gave _____ to the clerk.

   I got _____ cents in change.

Use with Lesson 8.5.

Date

1. Write the numbers.

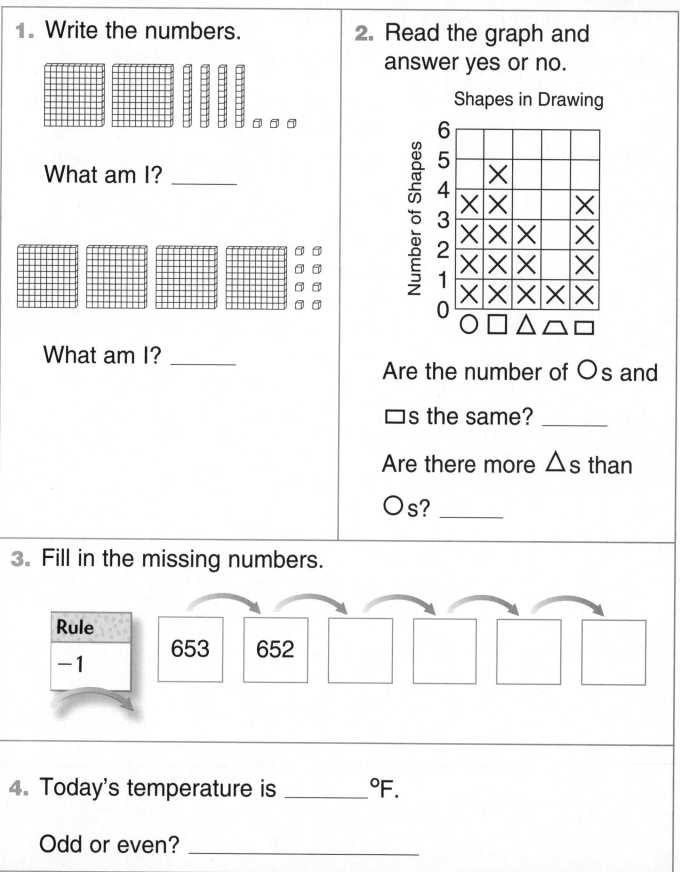

What am I? _____

What am I? _____

2. Read the graph and answer yes or no.

Shapes in Drawing

Are the number of ◯s and

▢s the same? _____

Are there more △s than

◯s? _____

3. Fill in the missing numbers.

Rule					
−1	653	652			

4. Today's temperature is _____ °F.

Odd or even? _____

Equal Shares

Show how you share your crackers.

1 cracker, 2 people	1 cracker, 4 people
Halves	**Fourths**
1 cracker, 3 people	2 crackers, 4 people
Thirds	

"What's My Rule?" Money Problems

Write the rule. Complete the table.

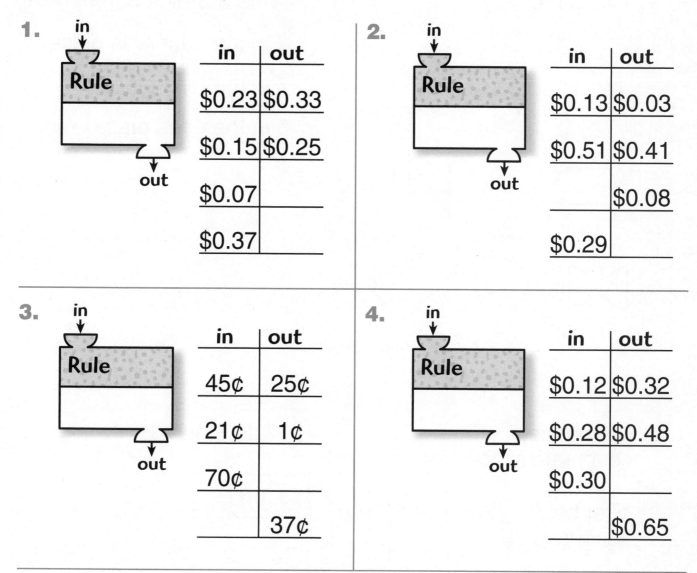

1.

in	out
$0.23	$0.33
$0.15	$0.25
$0.07	
$0.37	

2.

in	out
$0.13	$0.03
$0.51	$0.41
	$0.08
$0.29	

3.

in	out
45¢	25¢
21¢	1¢
70¢	
	37¢

4.

in	out
$0.12	$0.32
$0.28	$0.48
$0.30	
	$0.65

Make up your own.

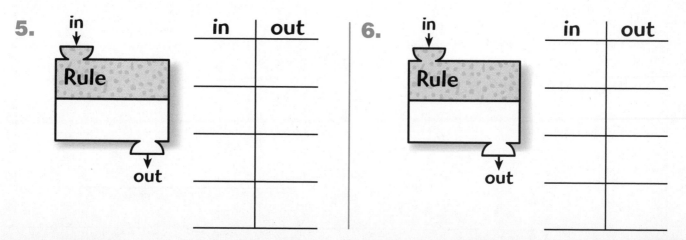

5.

in	out

6.

in	out

Math Boxes 8.6

1. Circle the polygons.

2. Write a 3-digit number with

6 in the hundreds place,

7 in the tens place,

8 in the ones place.

_____

3. Catherine:
Ⓠ Ⓠ Ⓝ Ⓓ Ⓓ Ⓓ $___.___

Stephen:
Ⓠ Ⓠ Ⓓ Ⓠ Ⓠ $___.___

Who has more?

_____

How much more?

_____

4. Write 5 names.

$1.00

Use with Lesson 8.6.

Equal Parts of Wholes

Which glass is half full? Circle it.

Which rectangles are divided into thirds? Circle them.

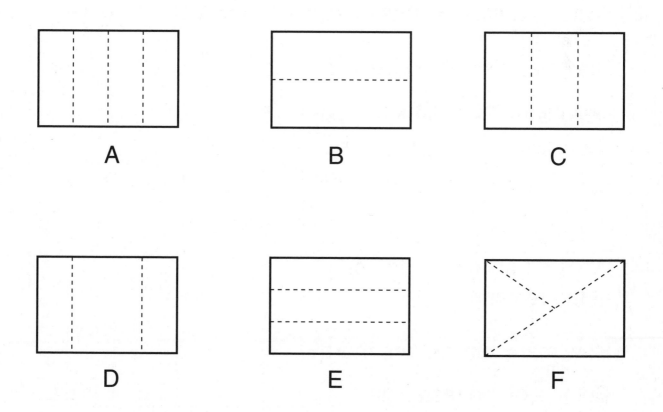

Fractions

1. Write a fraction in each part of the circle.

How many thirds are there? ____

Color $\frac{1}{3}$ of the circle.

2. Write a fraction in each part of the square.

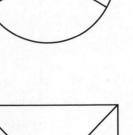

How many fourths are there? ____

Color $\frac{1}{4}$ of the square.

3. Write a fraction in each part of the hexagon.

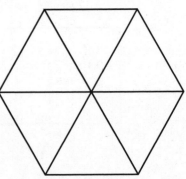

How many sixths are there? ____

Color $\frac{1}{6}$ of the hexagon.

4. Write a fraction in each part of the rectangle.

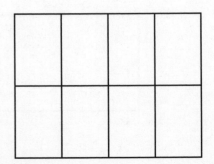

How many eighths are there? ____

Color $\frac{1}{8}$ of the rectangle.

Use with Lesson 8.7.

Math Boxes 8.7

1. Use your calculator. Count up by 50s.

<u>550</u>, <u>600</u>, <u>650</u>,

_____, _____, _____,

_____, _____, _____,

_____, _____

2. I bought a kite for $1.86.

I gave $2.00.

How much change did I get back?

_____

3. Divide each shape in half. Shade one half of each shape.

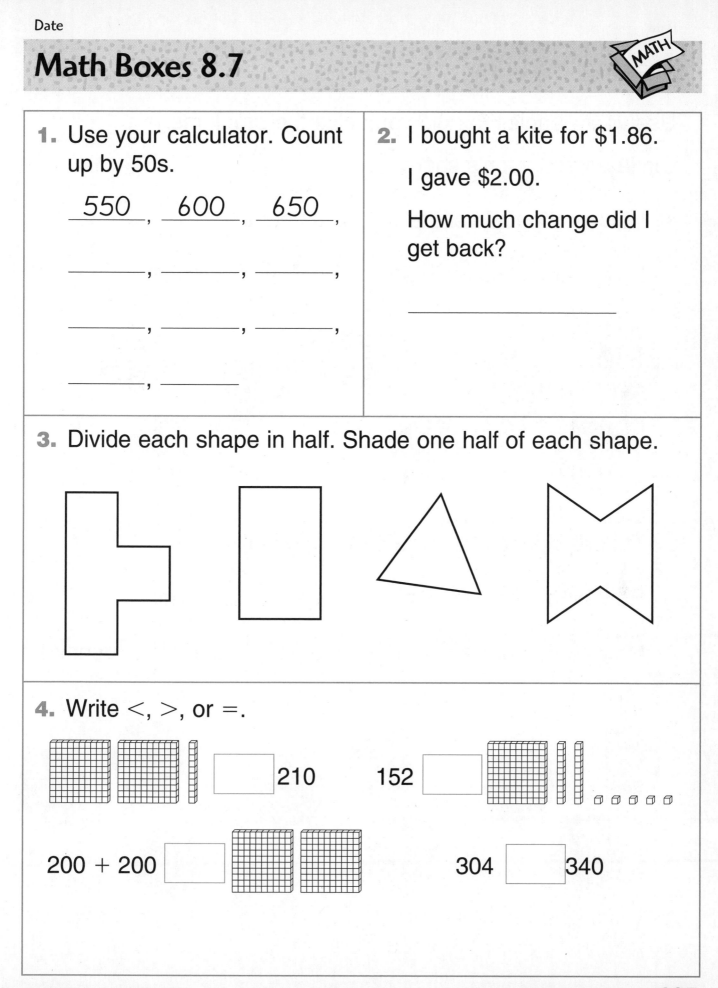

4. Write <, >, or =.

[base-ten blocks: 2 flats, 1 long] ☐ 210   152 ☐ [base-ten blocks: 1 flat, 2 longs, 5 cubes]

200 + 200 ☐ [base-ten blocks: 2 flats]   304 ☐ 340

Sharing Pennies

Use your pennies to help you solve the problems.

Circle each person's share.

1. Halves: 2 people share 8 pennies equally.

   How many does each person get? _____ pennies

2. Thirds: 3 people share 9 pennies equally.

   How many does each person get? _____ pennies

   How many pennies do 2 of the 3 people get? _____ pennies

Use with Lesson 8.8.

Sharing Pennies (cont.)

3. Fifths: 5 people share 15 pennies equally.

How many does each person get? _____ pennies

How many pennies do 3 of the 5 people get? _____ pennies

---

4. Fourths: 4 people share 20 pennies.

How many does each person get? _____ pennies

How many pennies do 2 of the 4 people get? _____ pennies

Date

Math Boxes 8.8

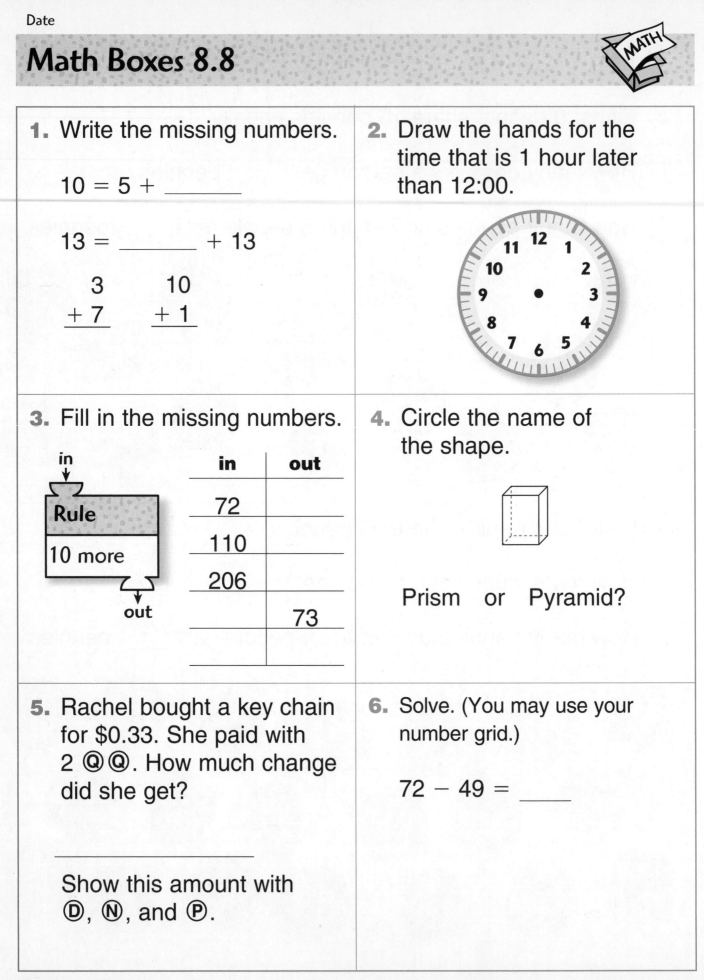

1. Write the missing numbers.

$10 = 5 +$ _____

$13 =$ _____ $+ 13$

$\begin{array}{r} 3 \\ + 7 \\ \hline \end{array}$ $\begin{array}{r} 10 \\ + 1 \\ \hline \end{array}$

2. Draw the hands for the time that is 1 hour later than 12:00.

3. Fill in the missing numbers.

in

Rule

10 more

out

in	out
72	
110	
206	
	73

4. Circle the name of the shape.

Prism   or   Pyramid?

5. Rachel bought a key chain for $0.33. She paid with 2 @ @. How much change did she get?

_____

Show this amount with Ⓓ, Ⓝ, and Ⓟ.

6. Solve. (You may use your number grid.)

$72 - 49 =$ _____

200 (two hundred)

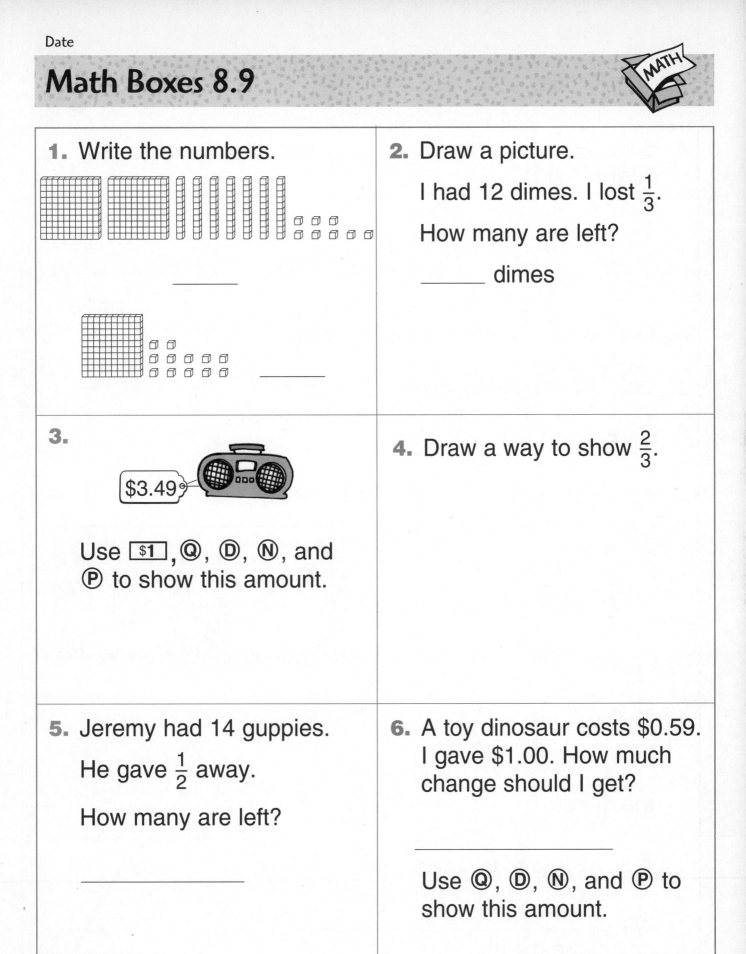

1. Write the numbers.

_____

_____

2. Draw a picture.

I had 12 dimes. I lost $\frac{1}{3}$.

How many are left?

_____ dimes

3.

$3.49

Use $1, Ⓠ, Ⓓ, Ⓝ, and Ⓟ to show this amount.

4. Draw a way to show $\frac{2}{3}$.

5. Jeremy had 14 guppies.

He gave $\frac{1}{2}$ away.

How many are left?

_____

6. A toy dinosaur costs $0.59. I gave $1.00. How much change should I get?

_____

Use Ⓠ, Ⓓ, Ⓝ, and Ⓟ to show this amount.

Date

MATH

1. Solve. (You may use your number grid.)

47 + 23 = _____

2. Use ⓠs. Show $2.00.

$\frac{1}{4}$ of $2.00 = $_____ . _____

3. Find the rule. Fill in the missing numbers.

| Rule | |

| | | 265 | 365 | | |

4. Measure the length of a tool-kit $1 bill.

_____cm   _____in.

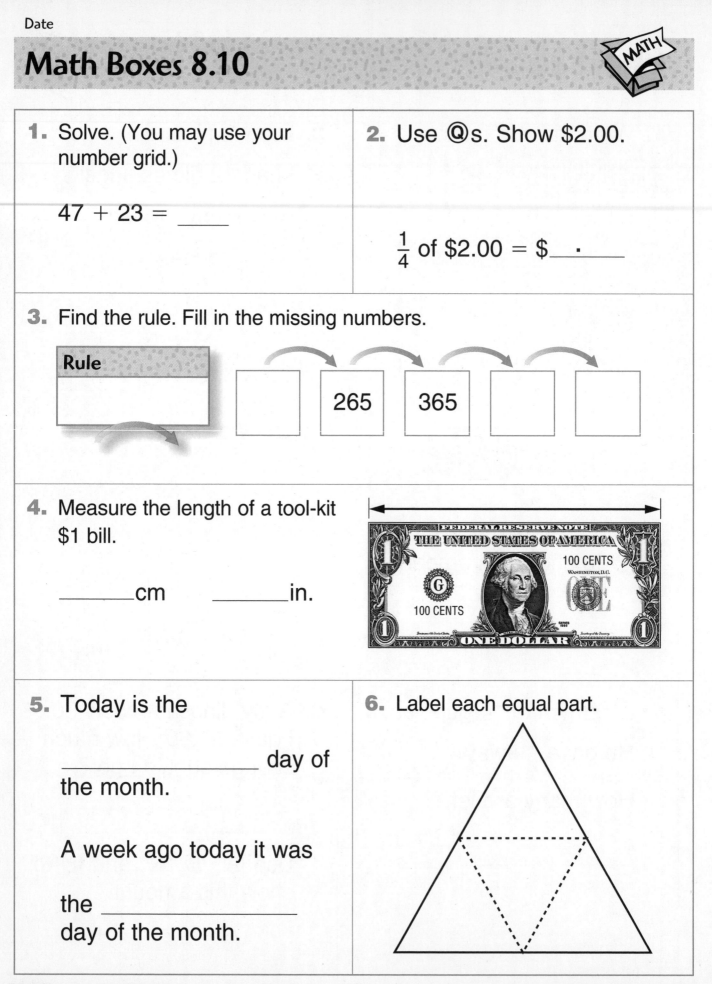

5. Today is the

_____ day of the month.

A week ago today it was

the _____
day of the month.

6. Label each equal part.

Number-Grid Hunt

0	10								90		
	9	19					69				109
	8			38				78			
	7										
	6								86		
	5	15				55					
	4			34							104
	3									93	
	2		22								
1											

The Smallest and the Largest

Use your 0–9 number cards. Choose two number cards. Make the smallest number you can. Make the largest number you can. Record the numbers.

	Digits Used	Smallest Number	Largest Number
Sample	5, 3	35	53

Choose three cards. Make the smallest number you can. Make the largest number you can. Record the numbers.

	Digits Used	Smallest Number	Largest Number
Sample	8, 0, 2	208	820

Use with Lesson 9.1.

1. Solve. You may use your number grid.

_____ = 82 − 30

2. For the number 102, write the number in the

hundreds place _____

tens place _____

ones place _____

3. Name or draw 2 objects shaped like rectangular prisms.

4. Fill in the oval for the correct time.

o 12:05

o 1:05

o 5:01

5. Rule

266   268   ☐   ☐   ☐

6. Complete the triangle. Write the fact family.

____ + ____ = ____   ____ − ____ = ____

____ + ____ = ____   ____ − ____ = ____

12

+, −

____   7

Number-Grid Game

Materials ☐ a number grid
☐ a die
☐ a game marker for each player

Players 2 or more

Take turns.

1. Put your marker at 0 on the number grid.

2. Roll the die. Use the table to see how many spaces to move the marker.

3. Continue. The winner is the first player to get to 110 or to get past 110.

Roll	Spaces
⚀	1 or 10
⚁	2 or 20
⚂	3
⚃	4
⚄	5
⚅	6

Use with Lesson 9.2.

Date

Color by Number

Use the code to color the picture.

7 = blue
8 = red
9 = green
10 = yellow

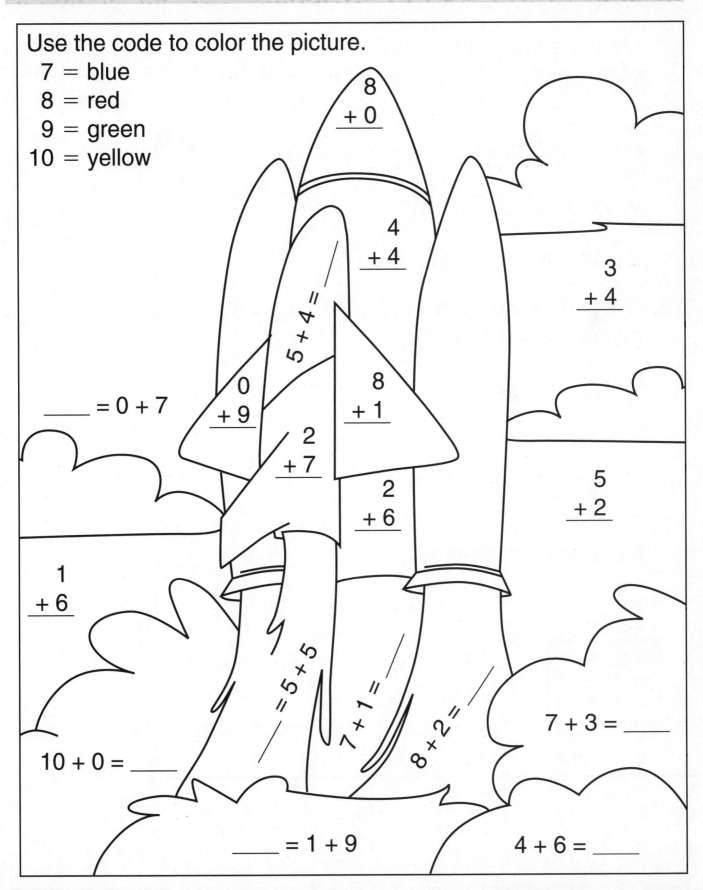

8
+ 0

4
+ 4

3
+ 4

5 + 4 =

0
+ 9

8
+ 1

____ = 0 + 7

2
+ 7

2
+ 6

5
+ 2

1
+ 6

= 5 + 5

7 + 1 =

8 + 2 =

7 + 3 = ____

10 + 0 = ____

____ = 1 + 9

4 + 6 = ____

Math Boxes 9.2

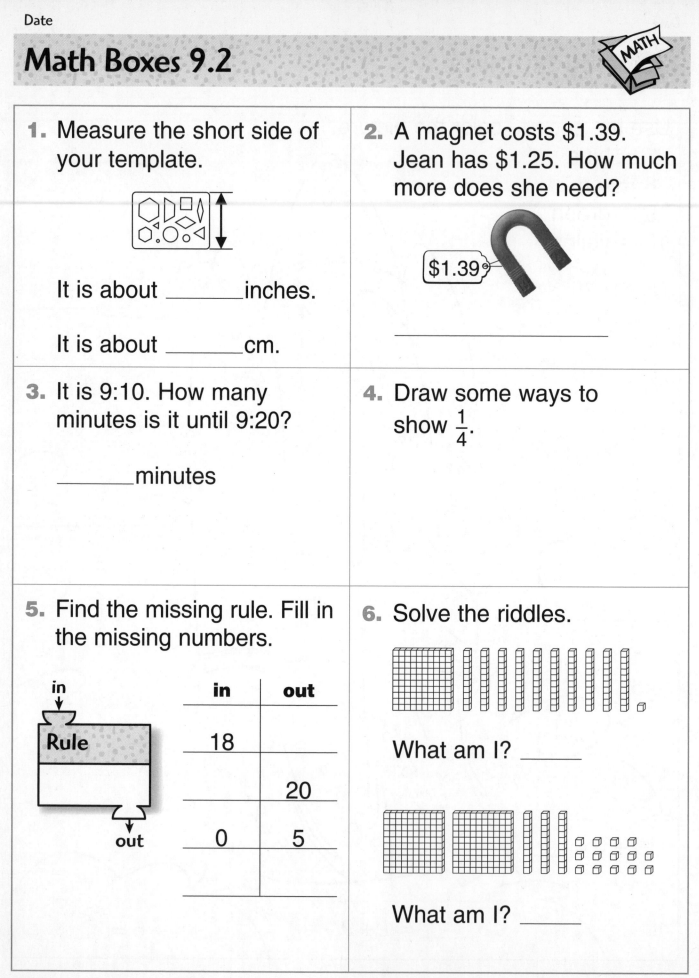

1. Measure the short side of your template.

   It is about _____ inches.

   It is about _____ cm.

2. A magnet costs $1.39. Jean has $1.25. How much more does she need?

   $1.39

   _____

3. It is 9:10. How many minutes is it until 9:20?

   _____ minutes

4. Draw some ways to show $\frac{1}{4}$.

5. Find the missing rule. Fill in the missing numbers.

   in

   Rule

   out

in	out
18	
	20
0	5

6. Solve the riddles.

   What am I? _____

   What am I? _____

Number-Grid Puzzles 1

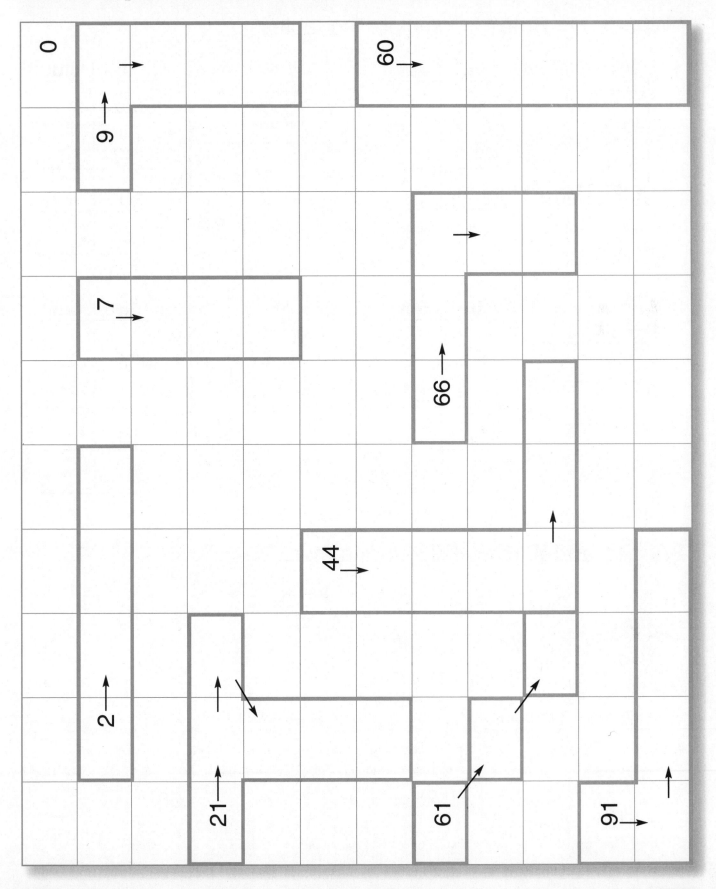

Using Rules to Solve Problems

"What's My Rule?" Complete the tables.

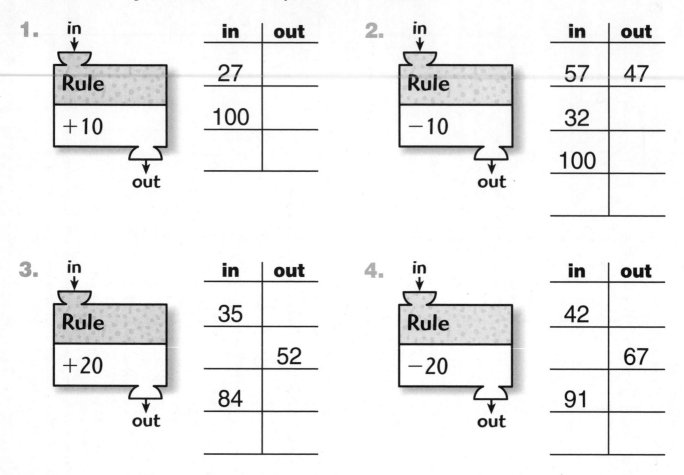

1.

in
Rule
+10
out

in	out
27	
100	

2.

in
Rule
−10
out

in	out
57	47
32	
100	

3.

in
Rule
+20
out

in	out
35	
	52
84	

4.

in
Rule
−20
out

in	out
42	
	67
91	

Frames-and-Arrows Fill in the frames.

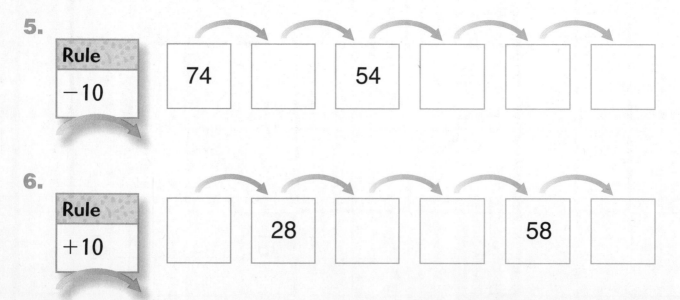

5.

Rule
−10

74 → ☐ → 54 → ☐ → ☐ → ☐

6.

Rule
+10

☐ → 28 → ☐ → ☐ → 58 → ☐

Date

Math Boxes 9.3

1. Complete the fact platter.

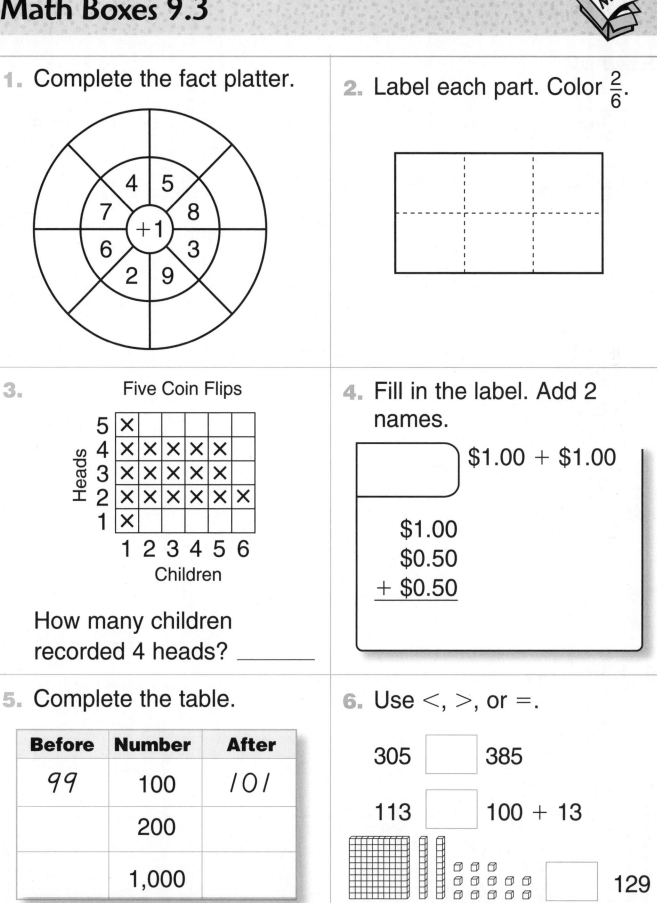

2. Label each part. Color $\frac{2}{6}$.

3. Five Coin Flips

How many children recorded 4 heads? _____

4. Fill in the label. Add 2 names.

$1.00 + $1.00

$1.00
$0.50
+ $0.50

5. Complete the table.

Before	Number	After
99	100	101
	200	
	1,000	

6. Use <, >, or =.

305 ☐ 385

113 ☐ 100 + 13

☐ 129

Silly Animal Stories

Use with Lesson 9.4.

Example

Unit
inches

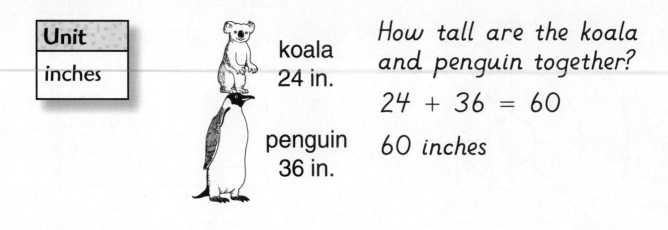

koala
24 in.

penguin
36 in.

How tall are the koala and penguin together?

24 + 36 = 60

60 inches

---

1. Silly Story

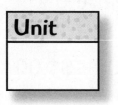

Unit

---

2. Silly Story

Unit

Number-Grid Puzzles 2

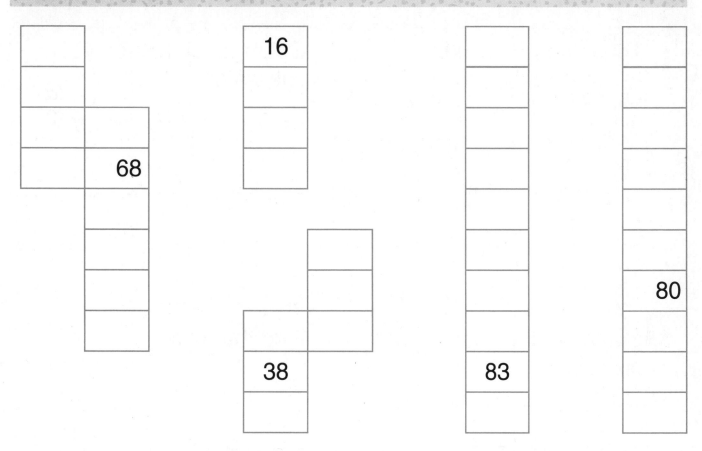

Make up your own pieces and paste them below.

Math Boxes 9.4

1. Ten dimes = $1.00.

One dime is _____ of a dollar.

2. Count up by 25s on your calculator.

_200_ , _225_ , _250_ ,

_____ , _____ , _____ ,

_____ , _____ , _____ ,

_____ , _____

3. Complete the number-grid puzzle.

42

4. Find the sum.

7 + 8 = _____

70 + 80 = _____

_____ = 9 + 8

_____ = 90 + 80

5. Record today's temperature.

_____ °F

Odd or even?

_____

6. Write the number that is 10 more.

_____

_____

Use with Lesson 9.4.

My Height Record

First Measurement

Date _____

Height: about _____ inches

A typical height for a first grader in my class is about

_____ inches.

Second Measurement

Date _____

Height: about _____ inches

A typical height for a first grader in my class is about

_____ inches.

The middle height for my class is about _____ inches.

Change to Height

I grew about _____ inches.

The typical growth in my class was about _____ inches.

Math Boxes 9.5

1. Show 83¢ in two ways.
Use Ⓠ, Ⓓ, Ⓝ, and Ⓟ.

2. Record the time.

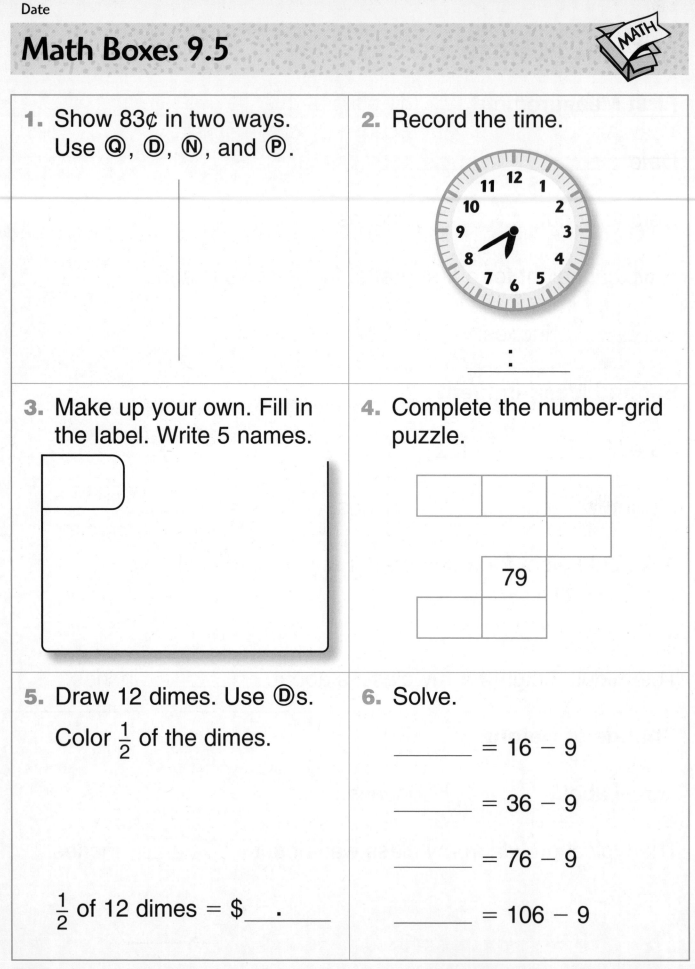

_____ : _____

3. Make up your own. Fill in the label. Write 5 names.

4. Complete the number-grid puzzle.

	79

5. Draw 12 dimes. Use Ⓓs.
Color $\frac{1}{2}$ of the dimes.

$\frac{1}{2}$ of 12 dimes = $_____ . _____

6. Solve.

_____ = 16 − 9

_____ = 36 − 9

_____ = 76 − 9

_____ = 106 − 9

Use with Lesson 9.5.

Pattern-Block Fractions

Use pattern blocks to divide each shape into equal parts.
Draw the parts with your Pattern-Block Template.
Color parts of the shapes.

1. Divide the rhombus into halves. Color $\frac{1}{2}$ of the rhombus.

2. Divide the trapezoid into thirds. Color $\frac{2}{3}$ of the trapezoid.

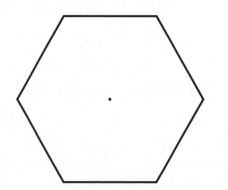

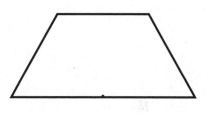

3. Divide the hexagon into halves. Color $\frac{2}{2}$ of the hexagon.

4. Divide the hexagon into thirds. Color $\frac{2}{3}$ of the hexagon.

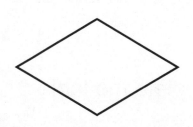

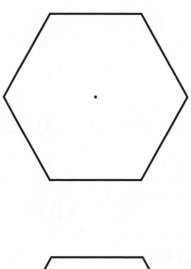

5. Divide the hexagon into sixths. Color $\frac{4}{6}$ of the hexagon.

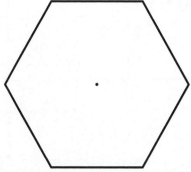

Math Boxes 9.6

---

1. You have $1.00. You buy pretzels that cost 65¢. How much change will you get?

_____¢

Show this amount.
Use Ⓠ, Ⓓ, Ⓝ, and Ⓟ.

2. Find the sums.

$2 + 8 + 4 =$ _____

$3 + 6 + 9 + 1 =$ _____

_____ $= 7 + 2 + 5$

---

3. Solve.

$\frac{1}{2}$ of 8 = _____

$\frac{1}{2}$ of 80 = _____

$\frac{1}{2}$ of 800 = _____

4. Complete the table.

Digits	Smallest	Largest
9, 3, 8		
4, 6, 1		

---

5. Fill in the missing numbers.

in

Rule

$-$ **$0.05**

out

in	out
$0.10	
	$0.10
$0.34	
	$0.60

6. Circle the polygons.

Fraction Strips

Use your fraction strips to help you answer the questions.

1. Which is more, $\frac{1}{2}$ of a 1-strip or $\frac{1}{4}$? _____ of a 1-strip

2. Which is more, $\frac{1}{4}$ or $\frac{1}{8}$? _____ of a 1-strip

3. Which is less, $\frac{1}{2}$ or $\frac{1}{8}$? _____ of a 1-strip

4. Which is less, $\frac{1}{2}$ or $\frac{1}{3}$? _____ of a 1-strip

5. Which is less, $\frac{1}{3}$ or $\frac{1}{6}$? _____ of a 1-strip

6. Which is more, $\frac{1}{3}$ or $\frac{1}{4}$? _____ of a 1-strip

7. Which is less, $\frac{1}{4}$ or $\frac{1}{6}$? _____ of a 1-strip

8. Which is more, $\frac{1}{2}$ or $\frac{1}{6}$? _____ of a 1-strip

Challenge

9. Which is more, $\frac{1}{2}$ or $\frac{2}{3}$? _____ of a 1-strip

1-strip

Date

Patterns and Pieces

1. Show counts by 2s with a /. Show counts by 4s with a \.

									0
1	2	3	4	5	6	7	8	9	10
11	12	13	14	15	16	17	18	19	20
21	22	23	24	25	26	27	28	29	30
31	32	33	34	35	36	37	38	39	40
41	42	43	44	45	46	47	48	49	50
51	52	53	54	55	56	57	58	59	60
61	62	63	64	65	66	67	68	69	70
71	72	73	74	75	76	77	78	79	80
81	82	83	84	85	86	87	88	89	90
91	92	93	94	95	96	97	98	99	100
101	102	103	104	105	106	107	108	109	110

Solve the number-grid puzzles.

2.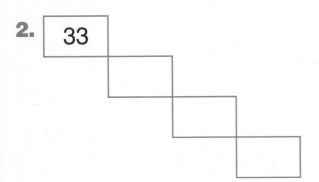

3.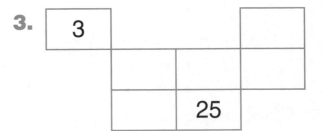

Make up your own.

4.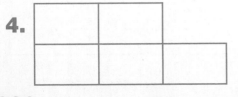

5.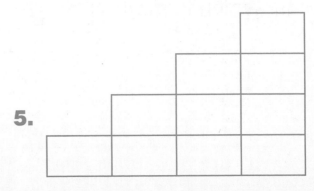

Use with Lesson 9.7.

Date

1. Draw the hands to show 11:45.

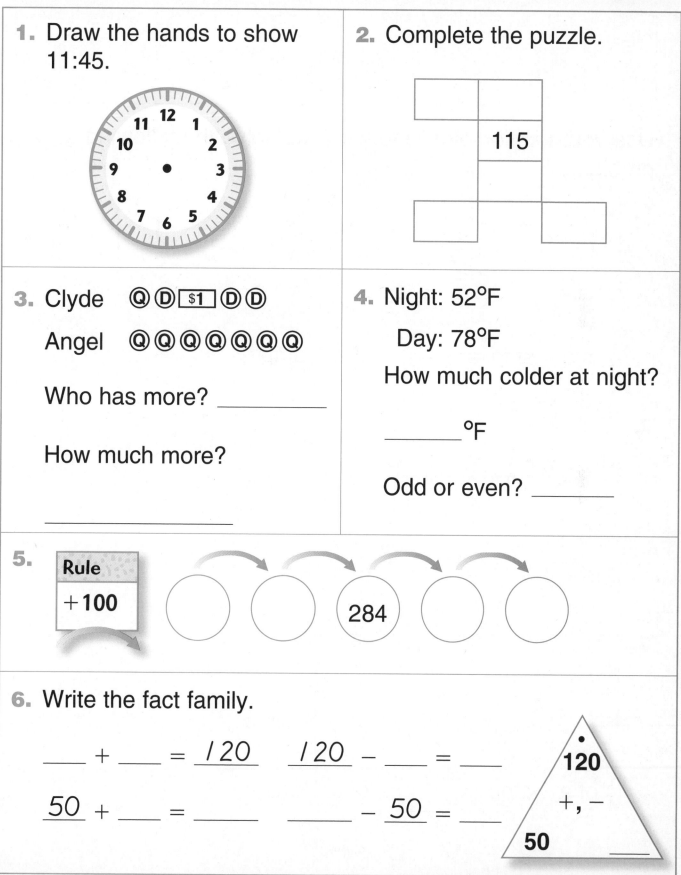

2. Complete the puzzle.

115

3. Clyde   Ⓠ Ⓓ $1 Ⓓ Ⓓ

   Angel   Ⓠ Ⓠ Ⓠ Ⓠ Ⓠ Ⓠ Ⓠ

   Who has more? _____

   How much more?

   _____

4. Night: 52°F

   Day: 78°F

   How much colder at night?

   _____°F

   Odd or even? _____

5.

Rule
+100

○ → ○ → 284 → ○ → ○

6. Write the fact family.

___ + ___ = 120   120 – ___ = ___

50 + ___ = _____   _____ – 50 = ___

120
+, –
50   ___

Many Names for Fractions

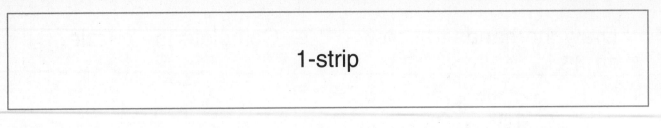

1-strip

Use your fraction pieces to help you solve the following problems.

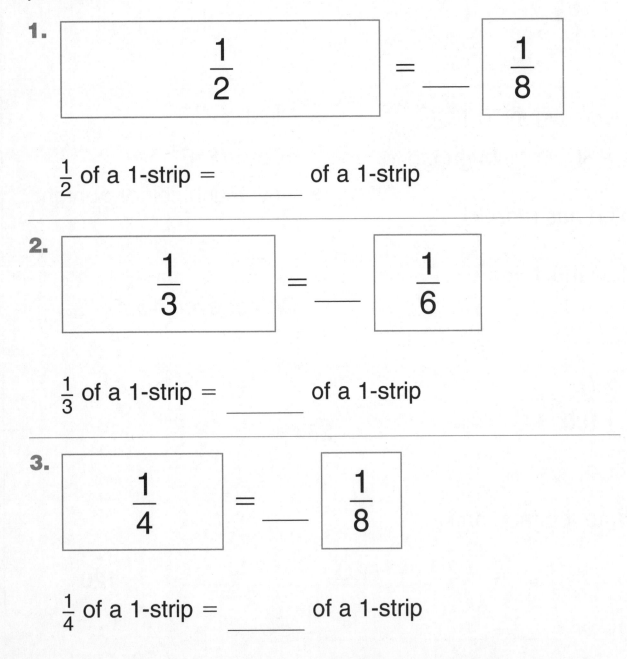

1.

$$\frac{1}{2} = \underline{\quad} \boxed{\frac{1}{8}}$$

$\frac{1}{2}$ of a 1-strip = _____ of a 1-strip

2.

$$\frac{1}{3} = \underline{\quad} \boxed{\frac{1}{6}}$$

$\frac{1}{3}$ of a 1-strip = _____ of a 1-strip

3.

$$\frac{1}{4} = \underline{\quad} \boxed{\frac{1}{8}}$$

$\frac{1}{4}$ of a 1-strip = _____ of a 1-strip

Many Names for Fractions (cont.)

4.

$\dfrac{1}{3}$	$\dfrac{1}{3}$	$=$ ___	$\dfrac{1}{6}$

$\dfrac{2}{3}$ of a 1-strip = _____ of a 1-strip

5.

$\dfrac{1}{4}$	$\dfrac{1}{4}$	$=$ ___	$\dfrac{1}{8}$

$\dfrac{2}{4}$ of a 1-strip = _____ of a 1-strip

6.

$\dfrac{1}{4}$	$\dfrac{1}{4}$	$\dfrac{1}{4}$	$=$ ___	$\dfrac{1}{8}$

$\dfrac{3}{4}$ of a 1-strip = _____ of a 1-strip

---

7. Cover the 1-strip with fraction pieces. Use at least 3 different sizes of fraction pieces. Draw line segments on the 1-strip to show which fraction pieces you used.

1-strip

1.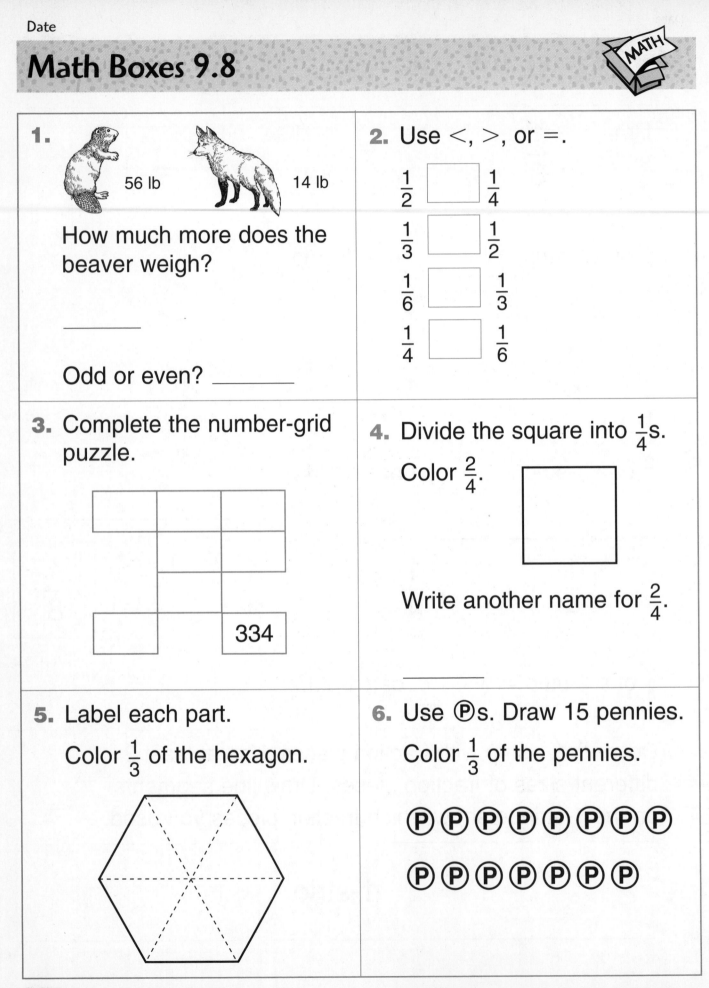

56 lb   14 lb

How much more does the beaver weigh?

_____

Odd or even? _____

2. Use <, >, or =.

$\frac{1}{2}$ ☐ $\frac{1}{4}$

$\frac{1}{3}$ ☐ $\frac{1}{2}$

$\frac{1}{6}$ ☐ $\frac{1}{3}$

$\frac{1}{4}$ ☐ $\frac{1}{6}$

3. Complete the number-grid puzzle.

334

4. Divide the square into $\frac{1}{4}$s.

Color $\frac{2}{4}$.

Write another name for $\frac{2}{4}$.

_____

5. Label each part.

Color $\frac{1}{3}$ of the hexagon.

6. Use Ⓟs. Draw 15 pennies.

Color $\frac{1}{3}$ of the pennies.

Ⓟ Ⓟ Ⓟ Ⓟ Ⓟ Ⓟ Ⓟ Ⓟ

Ⓟ Ⓟ Ⓟ Ⓟ Ⓟ Ⓟ Ⓟ

Math Boxes 9.9

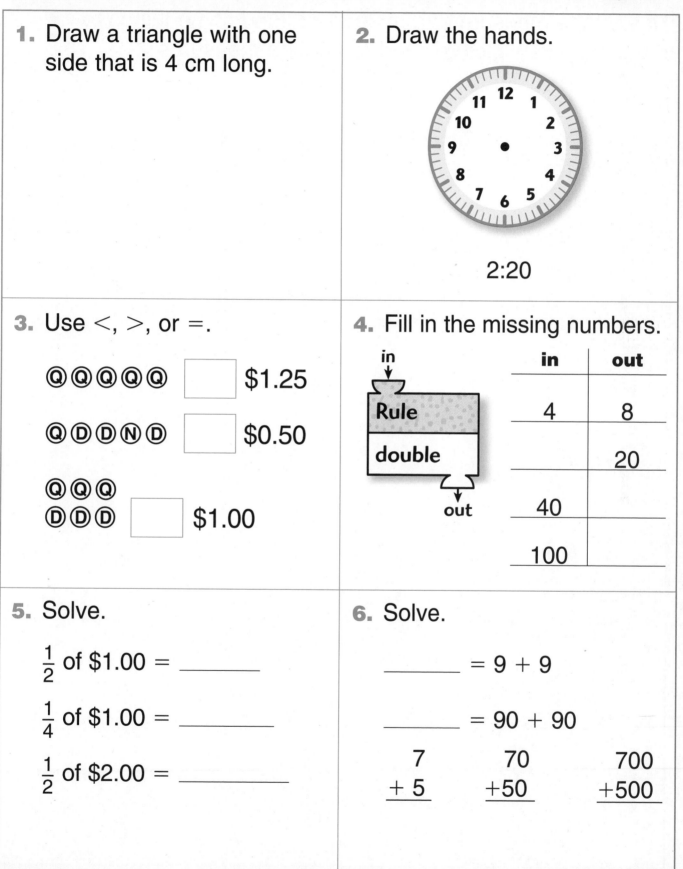

1. Draw a triangle with one side that is 4 cm long.

2. Draw the hands.

2:20

3. Use <, >, or =.

Ⓠ Ⓠ Ⓠ Ⓠ Ⓠ ☐ $1.25

Ⓠ Ⓓ Ⓓ Ⓝ Ⓓ ☐ $0.50

Ⓠ Ⓠ Ⓠ
Ⓓ Ⓓ Ⓓ ☐ $1.00

4. Fill in the missing numbers.

in
↓

Rule

double

out

in	out
4	8
	20
40	
100	

5. Solve.

$\frac{1}{2}$ of $1.00 = _____

$\frac{1}{4}$ of $1.00 = _____

$\frac{1}{2}$ of $2.00 = _____

6. Solve.

_____ = 9 + 9

_____ = 90 + 90

$\begin{array}{r} 7 \\ + 5 \\ \hline \end{array}$ $\begin{array}{r} 70 \\ +50 \\ \hline \end{array}$ $\begin{array}{r} 700 \\ +500 \\ \hline \end{array}$

Measurement Review

Use a tape measure or a meterstick. Find some things to measure. Draw pictures and record the measures.

_____ in.   _____ cm

_____ in.   _____ cm

_____ in.   _____ cm

Math Boxes 10.1

1. Write four even numbers with 9 in the hundreds place.

_____   _____

_____   _____

2. Draw the hands.

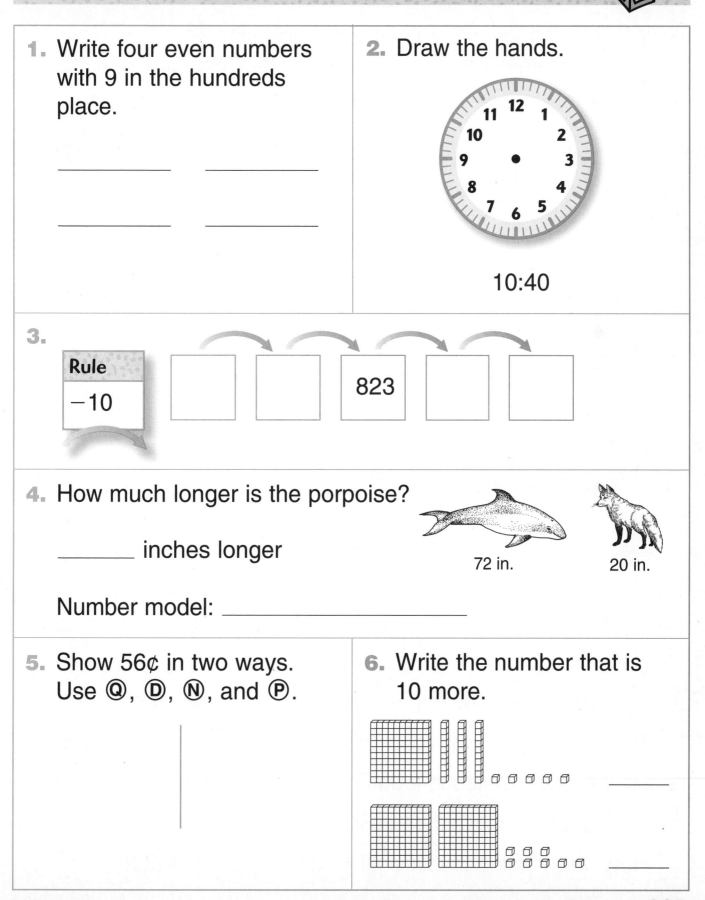

10:40

3.

Rule
−10

		823		

4. How much longer is the porpoise?

_____ inches longer

72 in.   20 in.

Number model: _____

5. Show 56¢ in two ways. Use Ⓠ, Ⓓ, Ⓝ, and Ⓟ.

6. Write the number that is 10 more.

_____

_____

Clock-Face Record

1. Ask a partner to show times on a tool-kit clock.
   Write the times and draw the hands for the times.

_____:_____   _____:_____   _____:_____

2. Write a time for each clock face. Draw the hands to match.

_____:_____   _____:_____   _____:_____

3. Set your tool-kit clock to 3:00.

   How many minutes is it until 3:30? _____ minutes

   Set your tool-kit clock to 1:30.

   How many minutes is it until 1:45? _____ minutes

   Set your tool-kit clock to 10:45.

   How many minutes is it until 11:15? _____ minutes

Fact-Extension Families

Cover the number under the dot to do addition
fact extensions. Cover one number at the bottom
of the Fact Triangle to do subtraction fact extensions.

Write the fact family for each triangle.

1.

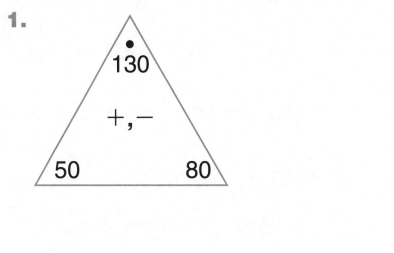

_50_ + _80_ = _130_

____ + ____ = ____

____ − ____ = ____

____ − ____ = ____

2.

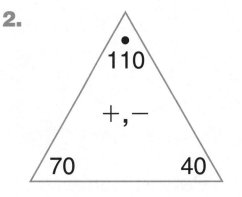

____ + ____ = ____

____ + ____ = ____

____ − ____ = ____

____ − ____ = ____

Math Boxes 10.2

---

1. $2.00 =

_____ pennies

_____ nickels

_____ dimes

_____ quarters

2. Write <, >, or =.

$10 + 23$ ⬜ 40

$18 + 5$ ⬜ $5 + 18$

30 ⬜ $51 - 20$

Half of 50 ⬜ 25

---

3. Divide the rectangle into fourths. Color $\frac{3}{4}$ of the rectangle.

4. Complete the table.

Digits	Smallest	Largest
5, 9, 0		
7, 1, 5		

---

5. Measure the height of your table or desk.

about _____ cm

6. Pets at Home

Children				
6	×			
5	×			
4	×			×
3	×	×		×
2	×	×	×	×
1	×	×	×	×
	Dog	Cat	Bird	Fish

How many pets in all? _____

Most popular pet: _____

Use with Lesson 10.2.

Number Stories and Number-Grid Pieces

1. Solve.

Tina has 2 sisters. Susan says she has 3 more sisters than Tina has.

Susan has _____ sisters.

Together, Tina and Susan have _____ sisters.

2. Solve.

My cat is _____ years older than my dog.

When my cat was 5 years old,

my dog was _____ years old.

my cat
(9 years old)

my dog
(7 years old)

Fill in the pieces of the number grids.

3.

	292	293	

4.

109	
	120

Buying from the Vending Machine

Pretend to buy items from the vending machine. Draw pictures or write the names of items you buy. Show the coins you use to pay for the items. Use Ⓟ, Ⓝ, Ⓓ, and Ⓠ. Write the total cost.

1.

2.

3.

4.

Show the cost of these items. Use Ⓟ, Ⓝ, Ⓓ, and Ⓠ. Write the total cost.

5.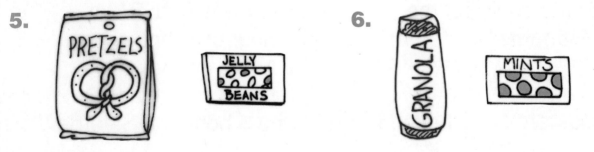

6.

Total cost: $_____

Total cost: $_____

Math Boxes 10.3

1. Write the number with

6 in the hundreds place,

0 in the tens place,

and 8 in the ones place.

_____

2. Draw the hands for 15 minutes later than 2:30.

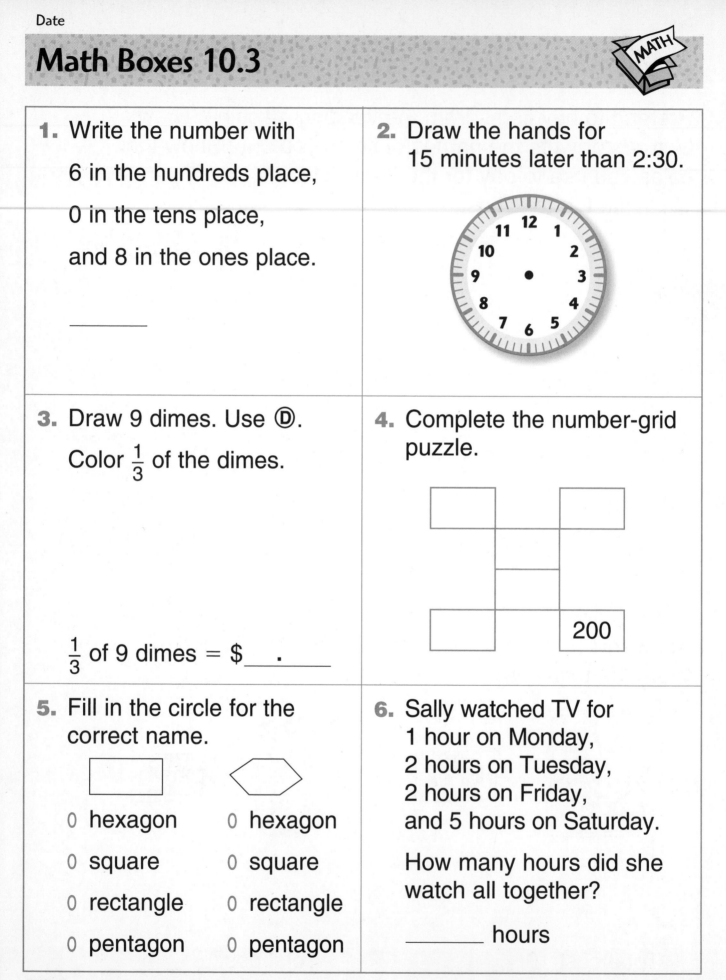

3. Draw 9 dimes. Use Ⓓ.

Color $\frac{1}{3}$ of the dimes.

$\frac{1}{3}$ of 9 dimes = \$____.____

4. Complete the number-grid puzzle.

200

5. Fill in the circle for the correct name.

o hexagon   o hexagon

o square   o square

o rectangle   o rectangle

o pentagon   o pentagon

6. Sally watched TV for
1 hour on Monday,
2 hours on Tuesday,
2 hours on Friday,
and 5 hours on Saturday.

How many hours did she watch all together?

_____ hours

Use with Lesson 10.3.

Tic-Tac-Toe Addition

Draw a straight line through 3 numbers if they equal the sum in the square. There is more than one combination for each sum.

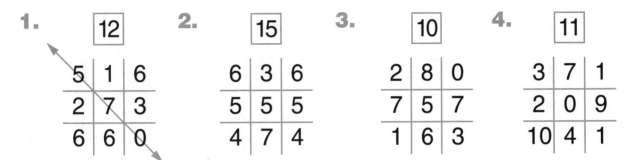

1.
12		
5	1	6
2	7	3
6	6	0

2.
15		
6	3	6
5	5	5
4	7	4

3.
10		
2	8	0
7	5	7
1	6	3

4.
11		
3	7	1
2	0	9
10	4	1

Numbers are missing in the puzzles below.

- Draw a straight line through 3 numbers if they equal the sum in the square.

- Fill in the missing numbers to make more combinations. Draw lines through the new combinations that equal the sum in the square.

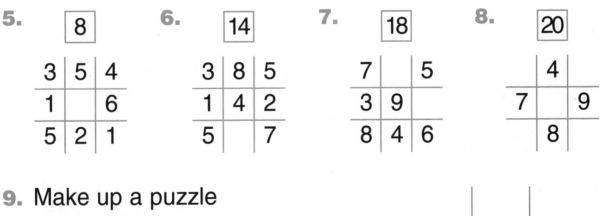

5.
8		
3	5	4
1		6
5	2	1

6.
14		
3	8	5
1	4	2
5		7

7.
18		
7		5
3	9	
8	4	6

8.
20		
	4	
7		9
	8	

9. Make up a puzzle of your own.

Date _____

1. Solve.

4 + 6 = _____

40 + 60 = _____

400 + 600 = _____

2. Complete the table.

Digits	Smallest	Largest
6, 2, 8		
5, 4, 1		

3. I buy 3 packs of gum at $0.25 each. I pay with $1. How much change will I get?

_____

Odd or even? _____

4. Write 5 names.

$\frac{1}{2}$

5. Write the fact family.

___ + ___ = __170__   __170__ − ___ = ___

___ + __80__ = ___   ___ − __80__ = ___

170
+, −
80   ___

6. Draw the next two figures.

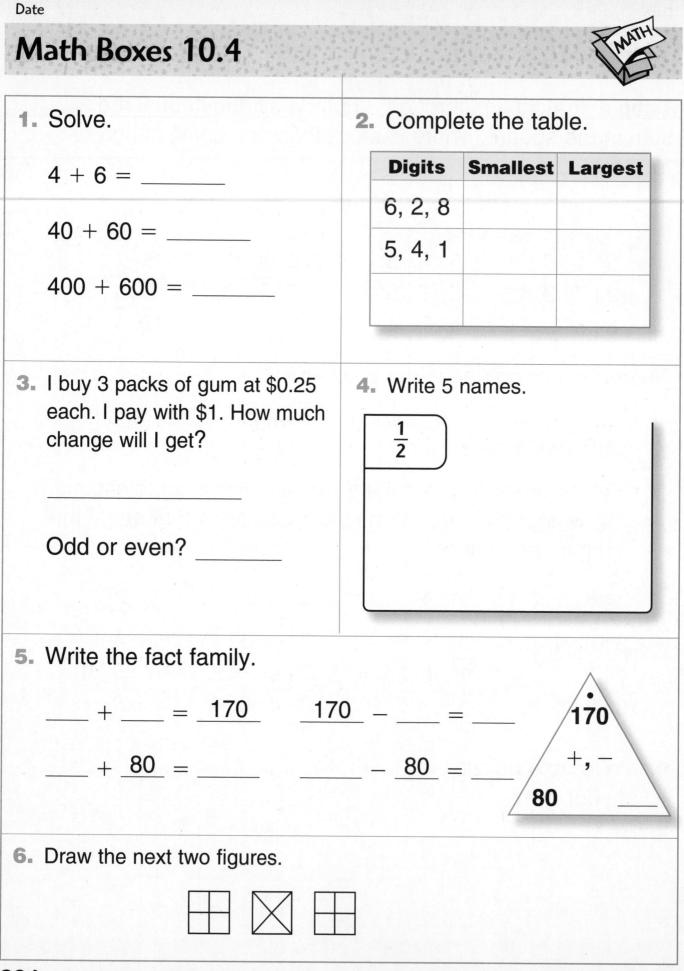

236 (two hundred thirty-six)

Use with Lesson 10.4.

Some Polygons

Triangles

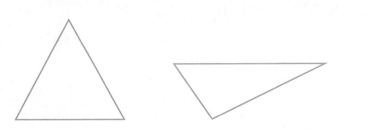

Quadrangles (Quadrilaterals)

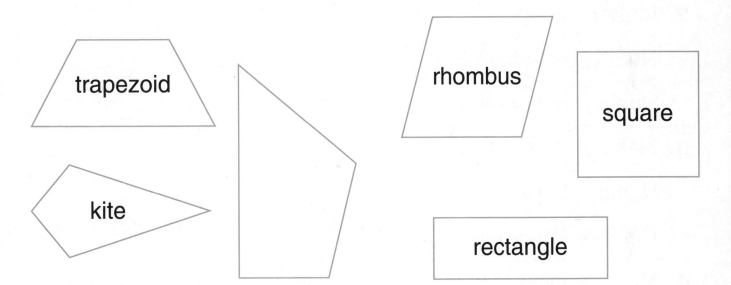

trapezoid

kite

rhombus

square

rectangle

Other Polygons

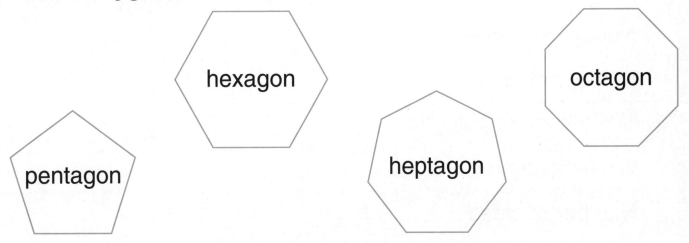

hexagon

octagon

pentagon

heptagon

Reviewing Polygons

Use straws and twist-ties to make the following polygons.
Draw the polygons. Record the number of corners and
sides for each polygon.

1. Make a square.

   Number of sides _____

   Number of corners _____

2. Make a triangle.

   Number of sides _____

   Number of corners _____

3. Make a hexagon.

   Number of sides _____

   Number of corners _____

4. Make a polygon of your choice.

   Write its name. _____

   Number of sides _____

   Number of corners _____

5. Make another polygon.

   Write its name. _____

   Number of sides _____

   Number of corners _____

Reviewing 3-Dimensional Shapes

Word Bank		
sphere	rectangular prism	pyramid
cube	cone	cylinder

Write the name of each 3-dimensional shape.

1.

_____

2.

_____

3.

_____

4.

_____

5.

_____

6.

_____

Challenge

7. Make a tetrahedron or a cube out of straws or twist-ties.

Five Regular Polyhedrons

The faces that make each shape are identical.

tetrahedron
4 faces

cube
6 faces

octahedron
8 faces

dodecahedron
12 faces

icosahedron
20 faces

Math Boxes 10.5

1. Record today's temperature.

_____°F

Odd or even?

_____

2. What time was it 10 minutes ago?

_____:_____

3. Show 65¢ in two ways.

Use Ⓠ, Ⓓ, Ⓝ, and Ⓟ.

4. Solve.

_____ + 7 = 9 + 1

6 + 6 = 3 + _____

5 + 4 = _____ + 2

_____ + 5 = 3 + 7

5. Tony is 15 years old. Maria is 7 years younger.

How old is Maria?

_____ years

Number model:

_____

6. Find the missing rule and numbers.

in

Rule

out

in	out
10	5
40	20
12	
	8

Use with Lesson 10.5.

Fact Extensions

1. $8 + 1 =$ _____

    $18 + 1 =$ _____

    $28 + 1 =$ _____

    $78 + 1 =$ _____

2. _____ $= 2 + 7$

    _____ $= 12 + 7$

    _____ $= 42 + 7$

    _____ $= 92 + 7$

3. $6 + 4 =$ _____

    $6 + 14 =$ _____

    $6 + 54 =$ _____

    $6 + 74 =$ _____

4. $9 - 5 =$ _____

    $19 - 5 =$ _____

    $39 - 5 =$ _____

    $89 - 5 =$ _____

5. _____ $= 10 - 2$

    _____ $= 30 - 2$

    _____ $= 50 - 2$

    _____ $= 70 - 2$

6. $16 - 8 =$ _____

    $26 - 8 =$ _____

    $56 - 8 =$ _____

    $86 - 8 =$ _____

7. 20

8. 70

9. 50

10. $\begin{array}{r} 300 \\ + 400 \\ \hline \end{array}$

Date

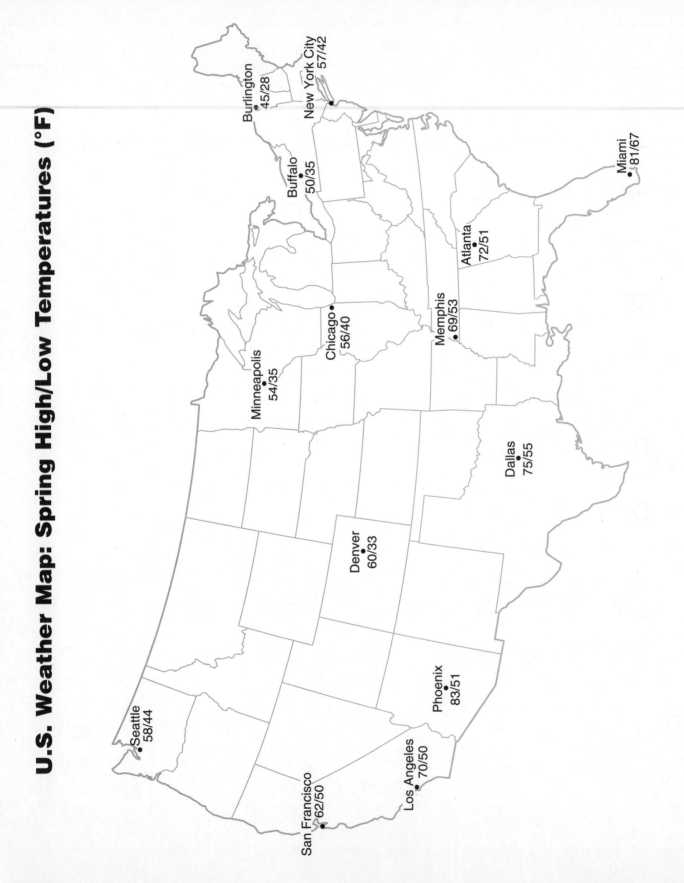

U.S. Weather Map: Spring High/Low Temperatures (°F)

Burlington
45/28

New York City
57/42

Buffalo
50/35

Miami
81/67

Atlanta
72/51

Memphis
69/53

Chicago
56/40

Minneapolis
54/35

Dallas
75/55

Denver
60/33

Phoenix
83/51

Seattle
58/44

Los Angeles
70/50

San Francisco
62/50

Temperature Chart

1.

City	Warmest Temperature	Coldest Temperature	Difference
_____	_____ °F	_____ °F	_____ °F
_____	_____ °F	_____ °F	_____ °F
_____	_____ °F	_____ °F	_____ °F
_____	_____ °F	_____ °F	_____ °F

2. Of your 4 cities,

_____ has the warmest temperature at _____ °F.

_____ has the coldest temperature at _____ °F.

The difference between these two temperatures is _____ °F.

3. On the map,

_____ has the warmest temperature at _____ °F.

_____ has the coldest temperature at _____ °F.

The difference between these two temperatures is _____ °F.

1. For the number 362, write the number

in the hundreds place. ____

in the tens place. ____

in the ones place. ____

2. How many fingers on 4 hands?

_____ fingers

Odd or even? _____

3. Yes or no?

1 hour > 60 minutes ____

12 months > 1 year ____

1 day = 24 hours ____

12 hours < $\frac{1}{2}$ day ____

4. You have $1.00. You buy a bag of cookies that costs 38¢. How much will you have left?

_____ ¢

Use Ⓠ, Ⓓ, Ⓝ, and Ⓟ to show this amount.

5. Solve.

$\frac{1}{2}$ of 10 = _____ $\frac{1}{2}$ of 20 = _____ $\frac{1}{2}$ of 100 = _____

6. Fill in the missing numbers.

Rule					
−100		650			

Date

1. How many toes are on 7 people?

_____ toes

Odd or even? _____

2. Write the number that is 10 less.

_____

_____

3. What time will it be in 10 minutes?

_____:_____

4. Count down by 1s on your calculator.

__*3*__ , __*2*__ , __*1*__ ,

_____ , _____ , _____ ,

_____ , _____ , _____

The smallest number is

_____

5. Complete the number-grid puzzle.

465

6. Regina Ⓓ Ⓓ Ⓠ Ⓝ Ⓠ $1

Salvo Ⓠ Ⓠ Ⓠ Ⓠ
Ⓠ Ⓠ Ⓠ Ⓠ

Who has more? _____

How much more?

_____

Equivalents and Abbreviations Table

Weight

kilogram: 1,000 g
pound: 16 oz
ton: 2,000 lb
1 ounce is about 30 g

<   *is less than*
>   *is more than*
=   *is equal to*
=   *is the same as*

Length

kilometer: 1,000 m
meter: 100 cm

foot: 12 in.
yard: 3 ft or 36 in.
mile: 5,280 ft or
               1,760 yd

10 cm is about 4 in.

Time

year:	365 or 366 days
year:	about 52 weeks
year:	12 months
month:	28, 29, 30, or 31 days
week:	7 days
day:	24 hours
hour:	60 minutes
minute:	60 seconds

Money

	1¢, or $0.01	Ⓟ
	5¢, or $0.05	Ⓝ
	10¢, or $0.10	Ⓓ
	25¢, or $0.25	Ⓠ
	100¢, or $1.00	$1

Abbreviations

kilometers	km
meters	m
centimeters	cm
miles	mi
feet	ft
yards	yd
inches	in.
tons	t
pounds	lb
ounces	oz
kilograms	kg
grams	g

Notes

Date

Notes

Notes

Notes

Fact Triangles 1

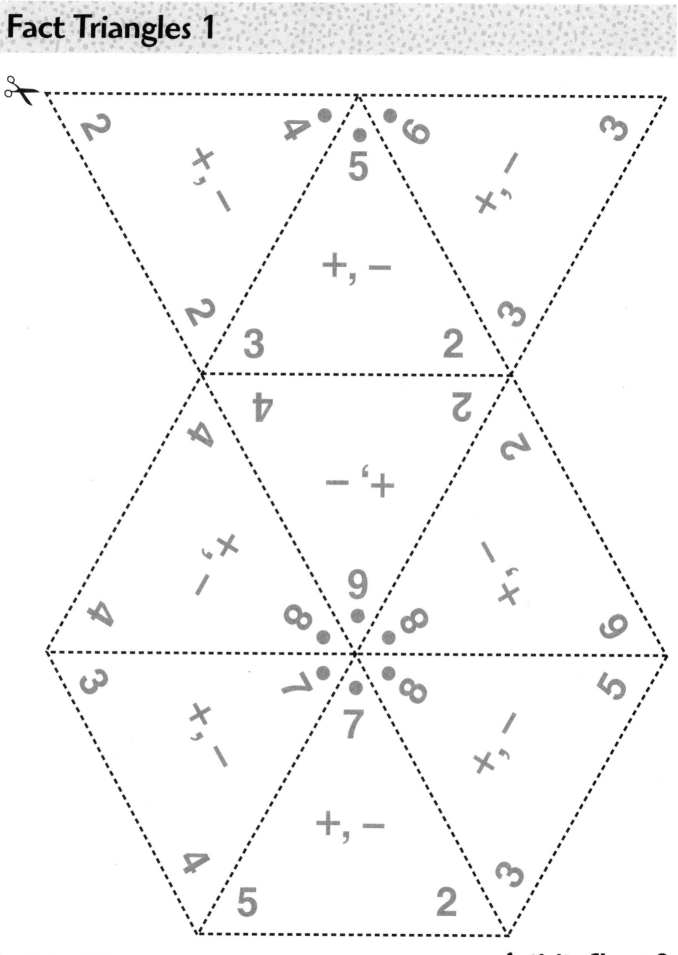

Fact Triangles 2

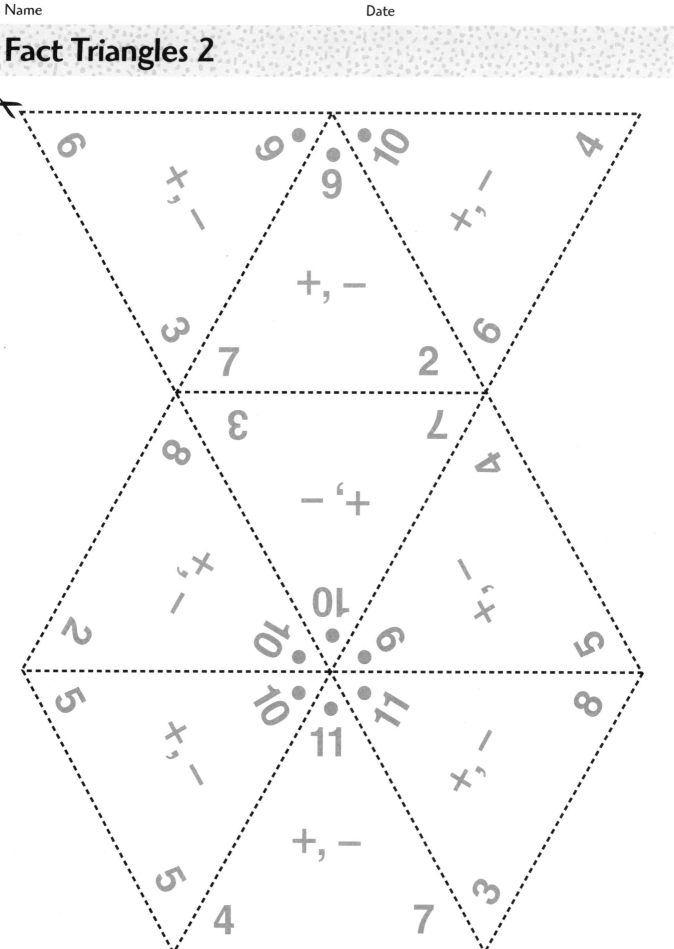

Fact Triangles 3

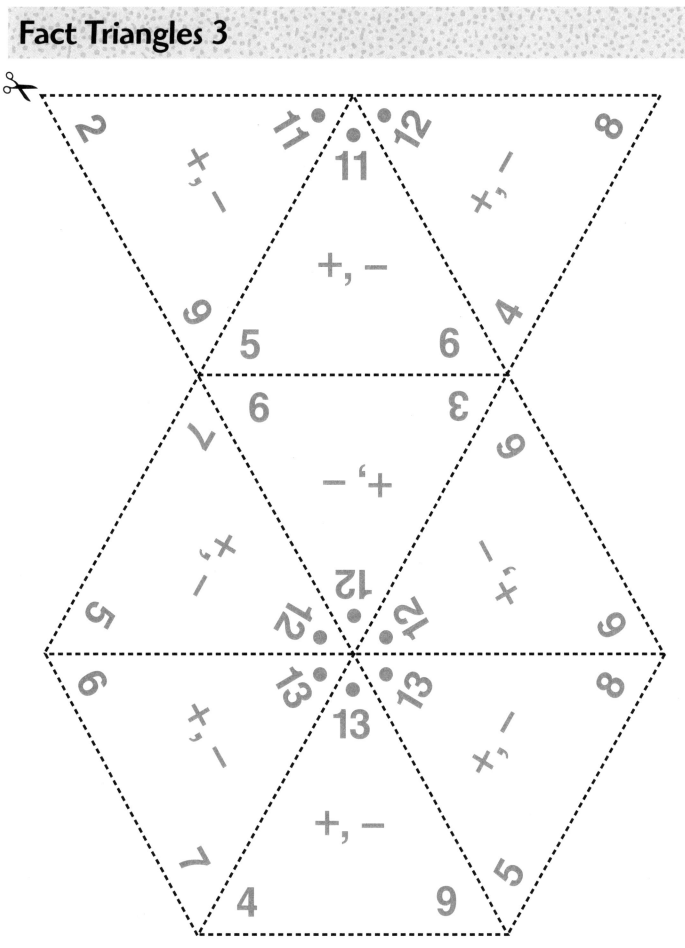

Activity Sheet 11

Fact Triangles 4

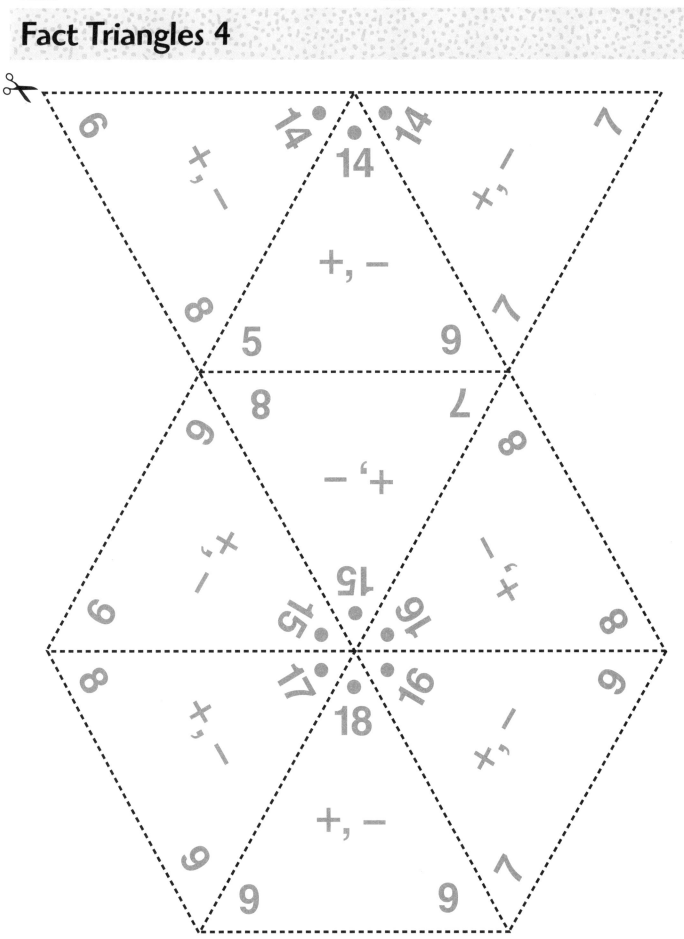

Number Grid

0	10	20	30	40	50	60	70	80	90	100	110
	9	19	29	39	49	59	69	79	89	99	109
	8	18	28	38	48	58	68	78	88	98	108
	7	17	27	37	47	57	67	77	87	97	107
	6	16	26	36	46	56	66	76	86	96	106
	5	15	25	35	45	55	65	75	85	95	105
	4	14	24	34	44	54	64	74	84	94	104
	3	13	23	33	43	53	63	73	83	93	103
	2	12	22	32	42	52	62	72	82	92	102
	1	11	21	31	41	51	61	71	81	91	101

Number Grid Shapes

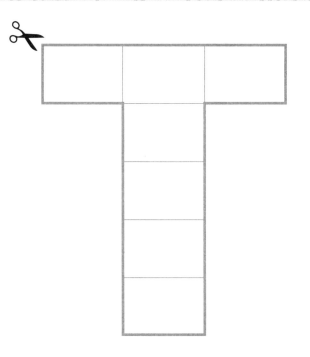

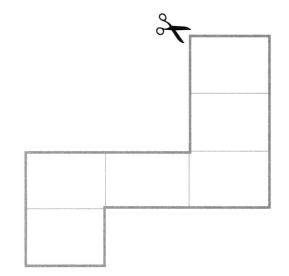

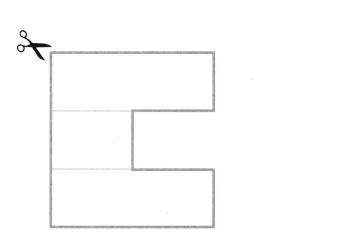

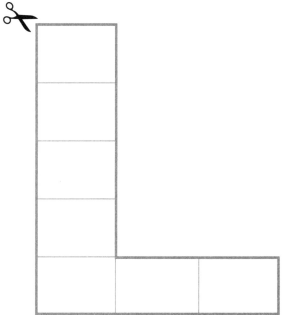